Generating a New Reality

From Autoencoders and Adversarial Networks to Deepfakes

Micheal Lanham

Apress®

Generating a New Reality: From Autoencoders and Adversarial Networks to Deepfakes

Micheal Lanham
Calgary, AB, Canada

ISBN-13 (pbk): 978-1-4842-7091-2   ISBN-13 (electronic): 978-1-4842-7092-9
https://doi.org/10.1007/978-1-4842-7092-9

Managing Director, Apress Media LLC: Welmoed Spahr
Acquisitions Editor: Celestin Suresh John
Development Editor: Laura Berendson
Coordinating Editor: Aditee Mirashi

Cover designed by eStudioCalamar

Cover image designed by Freepik (www.freepik.com)

Distributed to the book trade worldwide by Springer Science+Business Media New York, 1 New York Plaza, Suite 4600, New York, NY 10004-1562, USA. Phone 1-800-SPRINGER, fax (201) 348-4505, e-mail orders-ny@springer-sbm.com, or visit www.springeronline.com. Apress Media, LLC is a California LLC and the sole member (owner) is Springer Science + Business Media Finance Inc (SSBM Finance Inc). SSBM Finance Inc is a **Delaware** corporation.

For information on translations, please e-mail booktranslations@springernature.com; for reprint, paperback, or audio rights, please e-mail bookpermissions@springernature.com.

Apress titles may be purchased in bulk for academic, corporate, or promotional use. eBook versions and licenses are also available for most titles. For more information, reference our Print and eBook Bulk Sales web page at www.apress.com/bulk-sales.

Any source code or other supplementary material referenced by the author in this book is available to readers on GitHub via the book's product page, located at www.apress.com/978-1-4842-7091-2. For more detailed information, please visit www.apress.com/source-code.

Printed on acid-free paper

To my true loves: knowledge, my children, and Rhonda.

Table of Contents

About the Author

Micheal Lanham is a proven software and tech innovator with 20+ years of experience. During that time, he has developed a broad range of software applications in areas such as games, graphics, web, desktop, engineering, artificial intelligence, GIS, and machine learning applications for a variety of industries as an R&D developer. At the turn of the millennium, Micheal began working with neural networks and evolutionary algorithms in game development. He is an avid educator, and along with writing several books about game development, extended reality, and AI, he regularly teaches at meetups and other events. Micheal also likes to cook for his large family in his hometown of Calgary, Canada.

About the Technical Reviewer

Aneesh Chivukula is a technical expert and an analytics executive. He has strong academic research capacities in machine learning. He has developed innovative products with artificial intelligence. He brings thought leadership of technology trends in the enterprise solutions.

Aneesh has a doctorate of philosophy degree in data analytics and machine learning from the University of Technology Sydney, Australia. He holds a master of science degree by research in computer science and artificial intelligence from the International Institute of Information Technology Hyderabad, India.

Acknowledgments

This book, like many others, would not have been possible without the free exchange of knowledge provided in the AI/ML community, from countless researchers who tirelessly work on improving the field of artificial intelligence and generative modeling to the mass of AI enthusiasts who regularly produce open code repositories featuring new tools and a catalog of innovations.

I would like to thank and acknowledge all the contributions of those in the AI/ML field who work hard educating others. Many of the examples in this book have been collated from the numerous open-source repositories featuring deep learning and generative modeling. One such resource developed by Erik Linder-Norén, an ML engineer at Apple, inspired and contributed to several examples in this book's early chapters.

I would also like to thank you, the reader, for taking the opportunity to review this text and open your mind to new opportunities. It has always been a profound pleasure of mine watching that light bulb moment students experience when they first meld with a new concept. It's something I hope you will experience several times through the course of this book.

Lastly, as always, special thanks to my large family and friends. While I may not see all my nine children on a regular basis, they and their children always have a special place in my heart. I feel fortunate that my family supports my writing and continues to encourage new titles.

Introduction

We live in an era of fake news and uncertain reality. It's a world where reality has become blurred by digital wizardry artists and artificial intelligence practitioners. It's a digital reality now populated with fake news, images, and people. For many, the uncertainty and confusion are overwhelming. Yet, others, like yourself, search to embrace this new era of digital fakery to explore new opportunities.

This book takes an in-depth look at the technology that powers this new digital fake reality. The broad name for this technology or form of AI/ML is *generative modeling* (GM). It is a form of AI/ML modeling that looks to understand what something represents, as opposed to other forms of modeling that look to classify or predict something.

Generative modeling is not a new concept, but one that has emerged from the application of deep learning. The introduction of deep learning has launched the field into the mainstream. Unfortunately, not all mainstream use of GM is flattering or showcases the power of this diverse technology.

There is a real and speculated fear for most outside and inside the field of GM on what is possible. For many, the application of GM to produce fake anything is abhorrent and nonessential, but the broad applications GM introduces can benefit many industries across many tasks.

In this book, we begin with the assumption you have limited or little knowledge of deep learning and generative modeling. You have a basic knowledge of programming Python and applying data science, including the typical fundamental math knowledge in calculus, linear algebra, and statistics used in data science.

We will cover a wide range of GM techniques and applications in this book, starting with the fundamentals of building a deep learning network and then progressing to GM. Here is a brief overview of the chapters we will explore:

1. **The Basics of Deep Learning:** We begin by introducing the basic concepts of deep learning, autoencoders, and how to build simple models with PyTorch. The examples in this chapter demonstrate simple concepts we will apply throughout this book and should not be missed by newcomers.

2. **Unleashing Generative Adversarial Networks**: This chapter moves to the fundamentals of explaining the generative adversarial network (GAN) and how it can be used to generate new and novel content. Examples in this chapter explore the applications of GANs from generating fashion to faces.

3. **Exploring the Latent Space**: Fundamental to generative modeling is the concept of learning the latent or hidden representation of something. In this chapter, we explore how the latent space is defined and how we can better control it through hyperparameters, loss function, and network configuration.

4. **GANs, GANs, and More GANs**: This book explores several variations of GANs, and in this chapter we look at five forms that attempt to learn the latent space differently. We build on knowledge from previous chapters to explore key differences in the way GANs learn and generate content.

5. **Image to Image Content Generation**: This chapter covers the advanced application of GANs to enhance the generation of content by learning through understanding translations. The examples in this chapter focus on showcasing paired and unpaired image translation using a variety of powerful GANs.

6. **Residual Network GANs**: Throughout this book we will constantly struggle with the generative ability to produce diverse and realistic features. The GANs in this chapter all use residual networks to help identify and learn more realistic feature generation.

7. **Attention Is All We Need**: This chapter explores the attention mechanism introduced into deep learning through the application of natural language processing. Attention provides a unique capability to identify and map relevant features with other features. The examples in this chapter demonstrate the power of using an attention mechanism with a GAN.

8. **Advanced Generators**: This chapter dives into the deep end and explores the current class of best-performing GANs. The examples in this chapter work from several open-source repositories that showcase how far the field of GM has come in a short time.

9. **Deepfakes and Face Swapping**: In this chapter, we switch gears and explore the application of GM for producing deepfakes. Where this whole chapter is dedicated to showcasing the ease of which you can produce a deepfake freely available open-source desktop software.

10. **Cracking Deepfakes**: From creating deepfakes and fake content for most of the book, we move on to understanding how generated content can be detected. This chapter looks at the techniques and research currently being done to expose fake content. In the future, these tools will be critical to controlling the digital reality we embrace and understanding what is real.

This book covers a wide range of complex subjects presented in a practical hands-on and technically friendly manner. To get the most out of this book, it is recommended that you engage and work with several of the 40+ examples. All the examples in this book have been tested and run to completion using Google Colab, the recommended platform for this book. While some examples in this book may take up to days to train, most can be run in under an hour.

Thank you for taking your precious time to read this book and ideally expand your opportunities and understanding in the field of AI/ML. The journey you have chosen is nontrivial and will be filled with frustration and anguish. It is one that will also be filled with awe and wonder the first time you generate your first fake face.

CHAPTER 1

The Basics of Deep Learning

Throughout history mankind has often struggled with making sense of what is real and what reality means. From hunter gatherers to Greek philosophers and then to the Renaissance, our interpretation of reality has matured over time. What we once perceived as mysticism is now understood and regulated by much of science. Not more than 10 years ago we were on track to understanding the reality of the universe, or so we thought. Now, with the inception of AI, we are seeing new forms of reality spring up around us daily. New realities being manifested by this new wave of AI are made possible by *neural networks* and *deep learning*.

Deep learning and neural networks have been on the fringe of computer science for more than 50 years, and they have their own mystique associated with them. For many, the abstract concepts and mathematics of deep learning make them inaccessible. Mainstream science shunned deep learning and neural networks for years, and in many industries they are still off-limits. Yet, among all those hurdles, deep learning has become the brave new leader in AI and machine learning for the 21st century.

In this book, we look at how deep learning and neural networks work at a fundamental level. We will learn the inner workings of networks and what makes them tick. Then we will quickly move on to understanding how neural networks can be configured to generate their own content and reality. From there, we will progress through many other versions of deep learning content generation including swapping faces, enhancing old videos, and creating new realities.

For this chapter, we will start at the basics of deep learning and how to build neural networks for several typical machine learning tasks. We will look at how deep learning can perform regression and classification of data as well as understand internally the process of learning. Then we will move on to understanding how networks can be

© Micheal Lanham 2021
M. Lanham, *Generating a New Reality*, https://doi.org/10.1007/978-1-4842-7092-9_1

specialized to extract features in data with convolution. We will finish with building a full working image classifier using supervised deep learning.

As this is the first chapter, we will also cover several prerequisites and other helpful content to better guide your success through this book. Here is a summary of what we will cover in this chapter:

- Prerequisites

- Perceptrons

- Multilayer perceptrons

- PyTorch for deep learning

- Regression

- Classifying classes

This book will begin at the basics of data science, machine learning, and deep learning, but to be successful, be sure you meet most of the requirements in the next section.

Prerequisites

While many of the concepts regarding machine learning and deep learning should be taught at the high school level, in this book we will go way beyond the basic introduction of deep learning. Generating content with deep learning networks is an advanced endeavor that can be learned, but to be successful, it will be helpful if you meet most of the following prerequisites:

- **Interest in mathematics**: You don't need a degree in math, but you should have an interest in learning math concepts. Thankfully, most of the hard math is handled by the coding libraries we will use, but you still need to understand some key differences in math concepts. Deep learning and generative modeling use the following areas of mathematics:

  - **Linear algebra**, working with matrices and systems of equations

- **Statistics and probability**, understanding how descriptive statistics work and basic probability theory

- **Calculus**, understanding the basics of differentiation and how it can be used to understand the rate of change

- **Programming knowledge**: Ideally you have used and programmed with Python or another programming language. If you have no programming knowledge at all, you will want to pick up a course or textbook on Python. As part of your knowledge of programming, you may also want to take a closer look at the following libraries:

  - **NumPy**[1]: NumPy (pronounced "numb pie") is a library for manipulating arrays or tensors of numbers. It and the concepts it applies are fundamental to machine learning and deep learning. We will cover various uses of NumPy in this book, but it is suggested you study it further on your own as needed.

  - **PyTorch**[2]: This will be the basis for the deep learning projects in this book. It will be assumed you have little to no knowledge of PyTorch, but you may still want to learn more on your own what this impressive library has to offer.

  - **MatPlotLib**[3]: This module will be the foundation for much of the output we display in this book. There will be plenty of examples showing how it is used, but additional homework may be helpful.

- **Data science and/or machine learning**: It will be helpful if you have previously taken a data science course, one that covers the statistical methods used in machine learning and what aspects to be aware of when working with data.

---

[1]NumPy is an open source project at `http://numpy.org`.

[2]PyTorch is an open source project at `http://pytorch.org`.

[3]Matplotlib is an open source package heavily used with Python, available at `https://matplotlib.org/`.

- **Computer**: All the examples in this book are developed on the cloud, and while it is possible to use them with a mobile computing device, for best results it is suggested you use a computer.

    - Instructions have been provided in Appendix A for setting up and using the code examples on your local computer. This may be a consideration if you have a machine with an advanced GPU or need to run an example for more than 12 hours.

- **Time**: Generative modeling can be time-consuming. Some of the examples in this book may take hours and possibly days to run if you are up for it. In most cases, you will benefit more from running the example to completion, so please be patient.

- **Open to learn**: We will do our best to cover most of the material you need to use the exercises in this book. However, to fully understand some of these concepts, you may need to extend your learning outside this text. If your career is data science and machine learning or you want it to be, you likely already realize your path to learning will be continuous.

While it is highly recommended that you have some background in the prerequisites mentioned, you may still get by if you are willing to extend your knowledge as you read this book. There are many sources of text, blogs, and videos that you may find useful to help you fill in gaps in your knowledge. The primary prerequisites I ask you bring are an open mind and a willingness to learn.

In the next section, we jump into the foundation of neural networks, the perceptron.

The Perceptron

There is some debate, but most people recognize that the inspiration for neural networks was the brain, or, more specifically, the brain cell or neuron. Figure 1-1 shows the biological neuron over the top of a mathematical model called the *perceptron*. Frank Rosenblatt developed the basic perceptron model as far back as 1957. The model was later improved on to what is shown in the figure by Marvin Minsky in his book called *Perceptrons*. Unfortunately, the book was overly critical of the application of the perceptron for anything other than simple Boolean logic problems like XOR. Much of this criticism was unfounded as we later discovered, but the fallout of this critique is often blamed for the first AI winter.

An *AI winter* is when all research and development using AI is stopped or placed in storage. These winters are often brought on by some major roadblock that stops progress in the field. The first winter was brought on by Minsky's critique of the perceptron and his belief that it could solve the XOR problem only. There have been two AI winters thus far. The dates of these winters are up for debate and may vary by exact discipline.

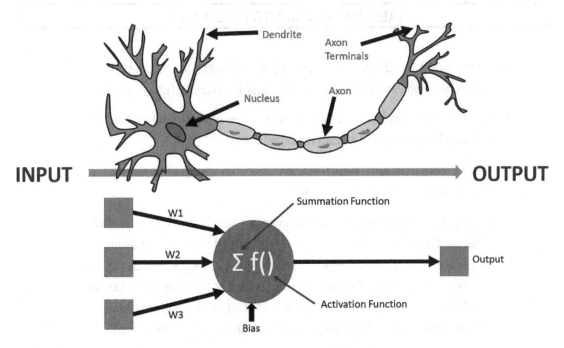

Figure 1-1. *A comparison of a biological neuron and the perceptron*

It is perhaps this association with the brain that causes some of the criticism with the perceptron and deep learning. This association also drives the mystique and uncertainty of neural networks. However, the perceptron itself is just a model of connectivity, and we may often refer to this type of learning as connectionism. If anything, the model of the perceptron only relates to a neuron in the way it connects and really nothing more. Actual neural brain function is far more complex and works nothing like a perceptron.

If we return to Figure 1-1 and the perceptron model, you can see how the system can take several inputs denoted by the boxes. These inputs are multiplied by a value we call a *weight* to weigh or adjust the strength of the input to the next stage. Before that, though, we have another input called a *bias*, with a value of 1.0, that we multiply by another weight. The bias allows the perceptron to offset the results. After the inputs and bias are all weighed/scaled, they are then collectively summed in the summation function.

The results of the summation function are then passed to an activation function. The purpose of the activation function may be to further scale, squish, or cut off the value to be output. Let's take a look at how a simple perceptron can be modeled in code in Exercise 1-1.

EXERCISE 1-1. CODING A PERCEPTRON

1. Open the GEN_1_XOR_perceptron.ipynb notebook from the project's GitHub site. If you are unsure on how to access the source, check Appendix B.

2. In the first code block of the notebook, we can see some imports for NumPy and Matplotlib. Matplotlib is used to display plots.

    ```
    import numpy as np
    import matplotlib.pyplot as plt
    ```

3. Scroll to the XOR problem code block, as shown here. This is where the data is set up; the data consists of the X and Y values that we want to train the perceptron on. The X values represent the inputs, and the Y values denote the desired output. We will often refer to Y as the label or the expected output. We use the numpy np module to create the lists of inputs to a tensor using np. array. At the bottom of this block, we output the shape of these tensors.

    ```
    X = np.array([[0,0],[0,1],[1,0],[1,1]])
    Y = np.array([0,1,1,0])

    print(X.shape)
    print(Y.shape)
    ```

4. The values we are using for this initial test problem are from the XOR truth table shown here:

Inputs		Outputs
X1	X2	Y
0	0	0
0	1	1
1	0	1
1	1	0

5. Scroll down and execute the following code block. This block uses the `matplotlib plt` module to output a 3D representation of the same truth table. We use array index slicing to display the first column of X, then Y, and finally the last column of X as the third dimension.

```
fig = plt.figure()
ax = fig.add_subplot(111, projection='3d')
ax.scatter(X[:,0], Y, X[:,1], c='r', marker='o')
```

6. Our first step in coding a perceptron is determining the number of inputs and creating the weights for those inputs. We will do that with the following code. In this code, you can see we get the number of inputs by taking the first value of the X.shape[1], which is 2. Then we randomly initialize the weights using np.random.rand and adding one input for the bias. Recall, the bias is a way the perceptron can offset a function.

```
no_of_inputs = X.shape[1]
weights = np.random.rand(no_of_inputs + 1)
print(weights.shape)
```

7. With the weights initialized to random values, we have a working perceptron.
    We can test this by running the next code block. In this block, we loop through
    the `inputs` called X and apply multiplication and addition using the dot
    product with the `np.dot` function. The output from this calculation yields the
    summation of the perceptron. The output of this code block will not mean
    anything yet since we still need to train the weights.

    ```
    for i in range(len(X)):
      inputs = X[i]
      print(inputs)
      summation = np.dot(inputs, weights[1:]) + weights[0]
      print(summation)
    ```

8. In the next code block is the training code to train the weights in the perceptron.
    We always train a perceptron or neural network iteratively over the data called
    in a cycle called an *epoch*. During each epoch or iteration, we will feed each
    sample of our data into the perceptron or network either singly or in batches.
    As each sample is fed, we compare the output of the summation function to the
    `label` or expected value, Y. The difference between the prediction and label is
    called the *loss*. Based on this loss, we can then adjust the weights based on a
    formula we will review in detail later. The entire training code is shown here:

    ```
    learning_rate = .1
    epochs = 100
    history = []
    for _ in range(epochs):
      for inputs, label in zip(X, Y):
        prediction = summation = np.dot(inputs, weights[1:]) + weights[0]
        loss = label - prediction
        history.append(loss*loss)
        print(f"loss = {loss*loss}")
        weights[1:] += learning_rate * loss * inputs
        weights[0] += learning_rate * loss
    ```

9. After the last code cell is run, run the last code cell, shown here, that generates
    a plot of the loss, as shown in Figure 1-2.

    ```
    plt.plot(history)
    ```

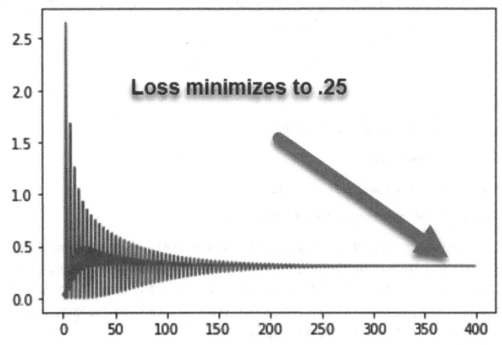

Figure 1-2. *Output of loss on XOR training of perceptron*

The results from this exercise are not so impressive. We were only able to obtain a minimized loss of .25. Feel free to continue running the example with more epochs or training cycles; however, the results won't get much better. This is the point Dr. Minsky was making in his book *Perceptrons*. A single perceptron or single layer of perceptrons is unable to solve the simple XOR problem. However, a single perceptron is able to solve some much harder problems.

Before we explore using the perceptron on a harder problem, let's revisit the learning lines of code from the previous example and understand how they work. For review, the learning lines of code are summarized here:

```
prediction = summation = np.dot(inputs, weights[1:]) + weights[0]
loss = label - prediction
...
weights[1:] += learning_rate * loss * inputs
weights[0] += learning_rate * loss
```

We already covered the summation/prediction function that uses `np.dot` to calculate. The loss is calculated by taking the difference from `label – prediction`. Then the weights are updated using the update function shown here:

$$W_i = W_i + \alpha * loss * input$$

where:

W_i = the weight that matches the input slot

α (alpha) = learning rate

loss = the difference from `label – prediction`

input = the input value for the input slot in the perceptron

This simple equation is what we use to update the weights during each pass of an input into the perceptron. The learning rate is used to scale the amount of update and is typically a value of .01, or 1 percent, or less. We want the learning rate to scale each update to a small amount; otherwise, each pass could cause the perceptron to over- and under-learn. The learning rate is the first in a class of variables we call *hyperparameters*.

Hyperparameters are a class of variables that we often need to tune manually. They are differentiated as hyperparameters since we refer to the internal weights as *parameters*.

The problem with a single perceptron or single layer of perceptrons is that they can solve a linear function only. The XOR problem is not a linear function. To solve XOR, we will need to introduce more than one layer of perceptron called a *multilayer perceptron*. Before we do that, though, let's revisit the perceptron and see what it is able to solve.

For the next exercise, we are going to look at a harder problem that can be solved with a linear method like the perceptron. The problem we will look at is solving a two-dimensional linear regression problem. Just 15 years ago, this class of problem would have been difficult to solve with typical regression methods. We will cover more about regression in a later section; for now let's jump into Exercise 1-2.

EXERCISE 1-2. LINEAR REGRESSION WITH A PERCEPTRON

1. Open the GEN_1_perceptron_class.ipynb notebook from the project's GitHub site. If you are unsure on how to access the source, check Appendix B.

2. This time we will run the linear regression problem code block to set the data, as shown here:

```
X = np.array([[1,2,3],[3,4,5],[5,6,7],[7,8,9],[9,8,7]])
Y = np.array([1,2,3,4,5])

print(X.shape)
print(Y.shape)
```

3. The next code block renders the input points on a graph:

```
fig = plt.figure()
ax = fig.add_subplot(111, projection='3d')
ax.scatter(X[:,0], X[:,1], X[:,2], c='r', marker='o')
```

4. In this case, we display just the input points in 3D on the plot shown in Figure 1-3. Our goal in this problem is to train the perceptron so that it can learn how to map those points to our output labels, Y.

```
<mpl_toolkits.mplot3d.art3d.Path3DCollection at 0x7f597e2abb00>
```

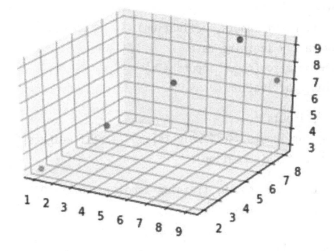

Figure 1-3. Input points plotted on 3D graph

5. We next move to the code section where we set up the parameters and hyperparameters. In this exercise, we have adjusted the hyperparameters, epochs and `learning_rate`. We decreased `learning_rate` to .01. Doing this effectively makes each update training pass or epoch less effective. However, in this case, the perceptron can learn to map those values much quicker than the XOR problem, so we will also reduce the number of epochs.

```
no_of_inputs = X.shape[1]
epochs = 50
learning_rate = .01
weights = np.random.rand(no_of_inputs + 1)
print(weights.shape)
```

6. For this exercise, we will introduce an activation function. An activation function scales the output for better input or prediction. In this example, we use a rectified linear function (ReLU). This function effectively negates output that is 0 or less and otherwise just passes the output linearly.

```
def relu_activation(sum):
  if sum > 0: return sum
  else: return 0
```

7. Next, we will embed the entire functionality of our perceptron into a Python class for better encapsulation and reuse. The following code is the combination of all our previous perceptron and setup code:

```
class Perceptron(object):
  def __init__(self, no_of_inputs, activation):
    self.learning_rate = learning_rate
    self.weights = np.zeros(no_of_inputs + 1)
    self.activation = activation

  def predict(self, inputs):
    summation = np.dot(inputs, self.weights[1:]) + self.weights[0]
    return self.activation(summation)

  def train(self, training_inputs, training_labels, epochs=100,
  learning_rate=0.01):
    history = []
    for _ in range(epochs):
```

```
        for inputs, label in zip(training_inputs, training_labels):
          prediction = self.predict(inputs)
          loss = (label - prediction)
          loss2 = loss*loss
          history.append(loss2)
          print(f"loss = {loss2}")
          self.weights[1:] += self.learning_rate * loss * inputs
          self.weights[0] += self.learning_rate * loss
      return history
```

8. We can instantiate and train this class with the following:

```
perceptron = Perceptron(no_of_inputs, relu_activation)
history = perceptron.train(X,Y, epochs=epochs)
```

9. Figure 1-4 shows the history output from the training function call and is a
   result of running the last group of cells. We can clearly see the loss is reduced
   to almost 0. This means our perceptron is able to predict and map the results
   given our inputs.

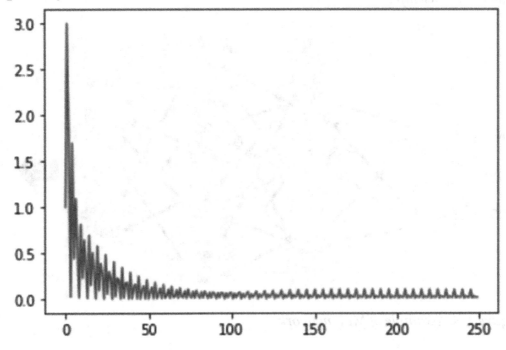

Figure 1-4. *Output loss of perceptron on linear regression problem*

You can see a noticeable wobble in the loss of the network in Figure 1-4. This wobble is caused in part by the learning rate, which is likely too high, and the way we are feeding the data into the network. We will look at how to resolve issues like this as we proceed through the book.

The results from this exercise were far more successful at mapping the inputs to expected outputs, even with a typically harder mathematical problem. Results like those we just witnessed are what kept the perceptron alive during the first cold AI winter. It wasn't until after this winter that we found the ability to stack perceptrons into layers could do far more and eventually solve the XOR problem. We will jump into the multilayer perceptron in the next section.

The Multilayer Perceptron

Fundamentally, the notion of stacking perceptrons into layers is not a difficult concept. Figure 1-5 demonstrates a three-layer multilayer perceptron (MLP). The top layer is called the *input layer*, the last layer the *output layer*, and the in-between layers the *middle* or *hidden layers*.

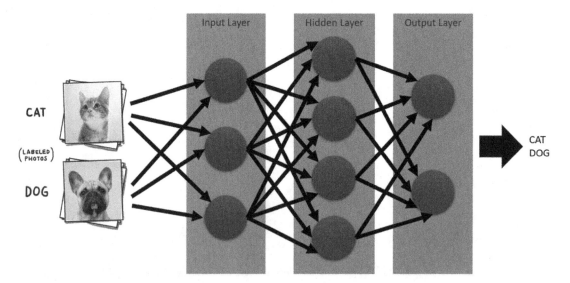

Figure 1-5. *Example of MLP network*

Figure 1-5 shows how we may feed images of cats and dogs to a network and have it classify the output. We will talk about how we can classify outputs later. Each node or circle in the figure represents a single perceptron, and each perceptron is fully connected to the successive layers in the network. The term we use for these types of networks is a *fully connected sequential network.*

The prediction of forward pass through the network runs the same as our perceptron, with the only difference that the output from the first layer becomes the input to the next, and so on. Calculating the output by passing an input into the network is called the *forward pass* or *prediction*. Computationally, through the use of the dot product function, the forward pass in DL is very efficient and one of the great strengths of neural networks.

If you recall from the previous section, the `np.dot` function we used did the summation of weights with the inputs. This function is optimized on a GPU to perform very quickly. So even if we had 1 million inputs (and yes, that is possible), the calculation could be done in one operation on a GPU.

The reason the `np.dot` function is optimized on a GPU is due to the advancement of computer 3D graphics. The dot product operation is quite common in graphics processing. In a sense, the development of games and graphics engines has been a big help for AI and deep learning.

While the forward pass or prediction step can run quickly, it is not exceedingly difficult to compute. Unfortunately, the opposite of training the updates or what we call the *backward pass* is not so easy. The problem we face when we stack perceptrons is that the simple update equation we applied before won't work across network layers.

The problem we encounter when updating the multiple layer networks is determining how much of the loss needs to be applied to not only which layer, but which perceptron in that layer. We can no longer just subtract the loss from the prediction and apply that value to a single weight. Instead, we need to calculate the impact of each weight applied to the resulting output or prediction.

To calculate how we can apply the loss to each weight, in each perceptron, and in each layer, we use calculus. Calculus allows us to determine the amount of loss to apply using differentiation. We use calculus to differentiate the forward or predict function along with the activation function. By differentiating these functions with respect to the weights, we can determine the amount of impact each weight contributes to the result.

Backpropagation

We call the update process or backward pass through network backpropagation since we are backpropagating the errors or loss to each weight. Figure 1-6 shows the backpropagation of error or loss back through the network. The figure also shows the equations to calculate this amount of error.

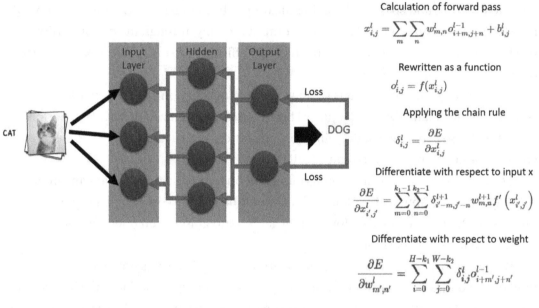

Figure 1-6. *Backpropagation explained*

The first equation in Figure 1-6 shows the calculation of the forward pass through the network. Moving to the next equation, we write this as a parameterized function. Following that, we apply the chain rule from calculus to differentiate the forward equation with respect to the input. By understanding how much each input affects the change, we can differentiate again this time with respect to the weights. The last equation shows how we calculate the change for each weight in the network.

Now, we don't have to worry about managing the mathematics or sorting out the equations to make this work. All deep learning libraries provide an automatic differentiation mechanism that does this for us. The critical bit to understand is how the last equation is used to push the loss back to each weight in the network. Output from this equation is a gradient describing the direction and amount of change. To perform the update, we reverse this gradient and scale it with the learning rate hyperparameter.

Since the output of our final calculation is a gradient, we use an optimization method to minimize or reduce this gradient. We call this method *gradient descent* since it intends to decrease the amount of gradient as the function minimizes the loss/error. You can picture how gradient descent works in Figure 1-7.

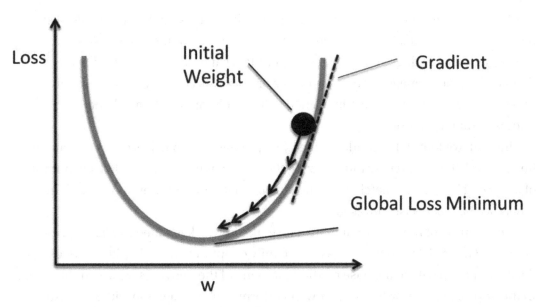

Figure 1-7. *Gradient descent explained*

In Figure 1-7, gradient descent is being used to modify the gradient of the weight to minimize the global loss of the forward pass function. The figure represents a simple 2D representation, but in most cases the dimensional space of the problem could well be in the hundreds or thousands. This means the landscape or surface of global minimum optimization for such a problem could have many hills and valleys of local minimums as well.

Stochastic Gradient Descent

Backpropagation and gradient descent were the discoveries that reawakened deep learning in the 1980s. Back then, though, they still had to work out the calculations manually. This was an impressive feat when you consider how complex or deep a network can get.

Over time, though, single gradient descent, doing the update after each input is shown to the network, was inefficient and as it turns out not optimal. What we found is that by batching data in groups, the process could be more efficient, and the results were not much different. We called this form of gradient descent *batch gradient descent* (BGD).

As deep learning matured, it was later found that consistent batches could skew results, especially if like datasets were clumped together. For instance, if we were classifying cats and dogs and our dataset had 1,000 cat images and 1,000 dog images, then feeding those images as batches of cats or dogs was contrary to doing small updates. A batch of 100 cats fed into a network would not provide any benefit to understanding what not to classify.

Instead, we found that randomizing the input data into batches of, say, cat and dog images worked far better. By randomizing the data, a network may be fed any number of cat or dog images in a batch, thus allowing the network to better understand the difference between a cat and dog.

The method of randomly batching data to the network is called *stochastic gradient descent* (SGD). SGD is now the basis for a family of optimization algorithms we use in deep learning. It will be the base method for many of the networks we use in this book. In the next section, we will look at how to perform backpropagation with stochastic gradient descent on a network with PyTorch.

PyTorch and Deep Learning

Up until now we have been using plain old Python with a little help from a friend called NumPy to construct some simple perceptrons. As you can likely appreciate now, the process of backpropagating loss through a network is not trivial and requires some additional help. Fortunately, there are numerous deep learning libraries available at our disposal, but not all of them are created equally.

For this book, we will use PyTorch, which is an open-source library that has become quite popular due to ease of use, performance, and ability to customize. It currently is the preferred library by most academic and cutting-edge AI researchers. This will benefit us when trying to build cutting-edge generative models later in this book.

PyTorch can allow us to build the simplest two-layer multilayer perceptron to several thousand-layer complex models. In Exercise 1-3, we will dive in and build our first MLP network in PyTorch to revisit and solve the XOR problem we failed at solving earlier.

EXERCISE 1-3. SOLVING XOR WITH MLP ON PYTORCH

1. Open the GEN_1_mlp_pytorch.ipynb notebook from the project's GitHub site. If you are unsure on how to access the source, check Appendix B.

2. Run the first block of code, which loads NumPy and several PyTorch (torch) modules we will work with.

    ```
    import numpy as np
    import torch
    import torch.nn as nn
    from torch.autograd import Variable
    import torch.nn.functional as F
    import torch.optim as optim
    ```

3. The next block of code contains the model in a class called XorNet. In this class, we create two neural network layers with the nn module using nn. linear. We call the layer fc; as a reminder, the layer is fully connected. The first fc1 layer takes 2 inputs and outputs to 10 neurons/perceptrons. The second fc2 layer takes 10 inputs and outputs to a single neuron. The forward function is where we can see how the input x is passed to the first fc1 layer and passed through a ReLU activation function with F.relu before being passed to the second fc2 layer.

    ```
    class XorNet(nn.Module):
      def __init__(self):
        super().__init__()
        self.fc1 = nn.Linear(2,10)
        self.fc2 = nn.Linear(10,1)
      def forward(self, x):
        x = F.relu(self.fc1(x))
        x = self.fc2(x)
        return x
    ```

19

4. The next block of code is where we instantiate the model of XorNet. After that, we create a loss function of type mean squared error (MSE). This loss calculation is the same as we saw earlier. After that, we create an optimizer of type Adam. The Adam optimizer is an improved version of SGD that provides for better scaling of the weight updates so that optimization can be more efficient.

5. From there, we move to setting up the data for the XOR problem and adjusting how this data is wrapped for consumption into PyTorch.

```
X = np.array([[0.,0.],[1.,1.],[0.,1.],[1.,0.]])
Y = np.array([0.,0.,1.,1.])

y_train_t = torch.from_numpy(Y).clone().reshape(-1, 1)
x_train_t = torch.from_numpy(X).clone()

history = []
```

6. Next, we move to training loops not unlike those we see before. The difference here is that we are batching the data. Since we have only four samples, we batch all the data. This is fed into the model to the forward pass or prediction. After that, the loss is calculated from the expected values in y_batch and those predicted y_pred using the loss function loss_fn. Then we zero the gradient of the optimizer. This is like a reset, with optimizer.zero_grad(). Apply the loss backward with loss.backward() and finally take another optimization step with optimizer.step().

```
for i in range(epochs):
  for batch_ind in range(4):
    x_batch = Variable(torch.Tensor(x_train_t.float()))
    y_batch = Variable(torch.Tensor(y_train_t.float()))
    y_pred = model(x_batch)
    loss = loss_fn(y_pred, y_batch)
    print(i, loss.data)
    optimizer.zero_grad()
    loss.backward()
    optimizer.step()
```

7. As the model is training, you will see the loss approaches zero quite quickly. Remember when we used a single perceptron model how our loss minimized at .25. This was due to the network not being able to fit the XOR function. By adding a second layer with 10 neurons/perceptrons, we are now able to solve XOR.

8. We can test how well the model is predicting by feeding a single input from the XOR truth table with the following:

```
v = Variable(torch.FloatTensor([1,0]))
model(v)
```

9. In the code you can see that we need to convert the input of 1,0 into a tensor and then into a torch variable with `Variable`. Then that value is fed into the model, and the prediction is displayed. Notice that the output of the prediction will be exactly 1.0.

By adding a second layer of 10 neurons to an MLP model, we were able to solve the XOR problem quite quickly and effectively using PyTorch. PyTorch provides the power to quickly build models with its neural network abstraction module nn. There are other lower ways to build networks in PyTorch, but we will stick with using nn for most of this book. In the next section, we look at how to exercise a PyTorch network for regression analysis.

Understanding Regression

Over the course of this chapter, we have used a form of teaching or learning called *supervised learning*. In supervised learning, our inputs have a known and expected output called a *label*. Mathematically, we often denote inputs as X and labels as Y shown in the familiar equation of a line, as shown here:

$$Y = mX + b$$

where:

m = slope or weight of the line

b = the offset or bias

Likewise, we can model the equation for a perceptron with the following:

$$Y = \sum_{i=1}^{n} w_i X_i + b$$

where:

n = the number of inputs

b = bias

In a supervised learning scenario, we are given the inputs X and the expected results or labels Y. The goal is to find the parameters of the equation, either slope and offset or weights and bias, that will produce the expected output. Typically, we solve these problems by performing regression analysis.

Regression analysis is when we iteratively modify the parameters of a model until we find a consistent solution for the set of all inputs. You can think of this analysis as mapping to or solving the function. For all deep learning and machine learning, our goal is often just solving an equation or function. Deep learning is an excellent equation solver when the equation is differentiable.

Since we use calculus to find the gradients, we can solve any equation if it is differentiable. That means the equation needs to be continuous at every point across the domain with no discontinuities or gaps.

The purpose of solving or learning an equation is to reuse that same equation for later inference of other unknown inputs to produce outputs. Thus, if we solve the equation of a line with a regression given a known set of data points and outputs, we can reuse that equation to plot or infer new data points.

Figure 1-8 shows the plot of data points X and Y with a line of regression passing through them. The equation of the line is shown in the legend. This plot was created with Microsoft Excel and uses the standard linear regression tools to generate the plot with a trendline. A trendline in Excel vernacular is a line calculated using regression to solve the line that best fits the points.

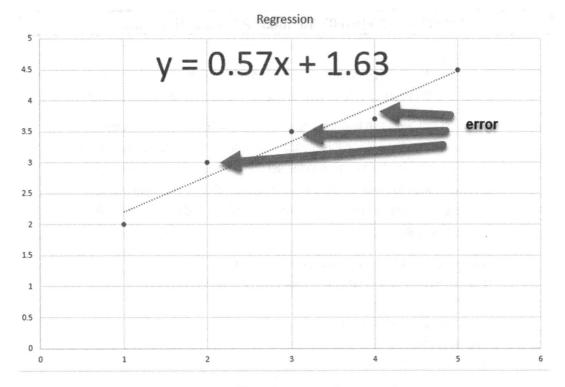

Figure 1-8. Trendline equation of line from Excel

The line shown on the plot can be used to further infer new data points given some arbitrary X value. For instance, if we wanted to determine what the value of Y would be when X is 0, the results would be 1.63, which so happens to the offset of intercept of the line. How accurate this line prediction is depending on the quantity and quality of the data.

When we run a deep learning model, we are essentially doing the same thing: solving for some unknown equation that will fit the input data. In future chapters, this process may become blurred, but it is important to understand that in all cases we are just solving for a function or equation with deep learning.

In data science we often denote simple regression as finding an explicit value. What you may consider the opposite of regression is *classification*. In classification, we don't find an explicit value but instead the class the value belongs to. We will explore more about classification later in this chapter. For now, though, let's jump into another exercise and show how we can perform more complex regression with a deep learning network. In Exercise 1-4, we will use the Boston housing value dataset to predict the value of a house.

```
┌─────────────────────────────────────────────────────────────────┐
│            EXERCISE 1-4. PREDICTING HOUSE PRICES WITH PYTORCH      │
└─────────────────────────────────────────────────────────────────┘
```

1. Open the GEN_1_regression_pytorch.ipynb notebook from the project's
    GitHub site. If you are unsure on how to access the source, check Appendix B.

2. In the first cell, we again load the imports and this time add a new module
    called sklearn. We will use this module to download the Boston housing data
    as well as break the data into a training set and a test set. In most cases, we
    will always want to break our data into a training/test split, where on average
    80 percent or the original data will be used for training and the remaining 20
    percent to test the model after.

    ```
    from sklearn.datasets import load_boston
    from sklearn.model_selection import train_test_split
    import numpy  as np
    import matplotlib.pyplot as plt
    import torch
    import torch.nn as nn
    ```

3. Next, we will load the dataset with the code and split it up into inputs (X) and
    labels (y).

    ```
    boston = load_boston()
    X,y   = (boston.data, boston.target)
    boston.data[:2]
    inputs = X.shape[1]
    ```

4. The data is loaded, and now we will split it into the training and test split
    using the train_test_split function. Notice how we set test_size to .2,
    meaning 20 percent of the data will be test.

    ```
    X_train, X_test, y_train, y_test = train_test_split(X, y, test_
    size=0.2, random_state=0)
    num_train = X_train.shape[0]
    X_train[:2], y_train[:2]
    num_train
    ```

5. This time we are going to create a sequential model with four layers. The first
    layer will take the size of the dataset as inputs and output to 50 neurons. It
    continues to the next layer with 50 and 50 and so on, finishing at the output
    layer of one neuron. Then we create the loss and optimizer functions. Notice the
    difference in activation functions between the layers and the output layer. The
    last layer uses a `sigmoid` function, which outputs a value from 0 to 1.

```
torch.set_default_dtype(torch.float64)
net = nn.Sequential(
    nn.Linear(inputs, 50, bias = True), nn.ReLU(),
    nn.Linear(50, 50, bias = True), nn.ReLU(),
    nn.Linear(50, 50, bias = True), nn.Sigmoid(),
    nn.Linear(50, 1)
)
loss_fn = nn.MSELoss()
optimizer = torch.optim.Adam(net.parameters(), lr = .001)
```

6. We can next move on to preparing the data for training. For this example, we
    need to reshape the labels from row to column order. In this block of code, we
    set up tensors for both the training and testing datasets.

```
num_epochs = 8000
y_train_t = torch.from_numpy(y_train).clone().reshape(-1, 1)
x_train_t = torch.from_numpy(X_train).clone()
y_test_t = torch.from_numpy(y_test).clone().reshape(-1, 1)
x_test_t = torch.from_numpy(X_test).clone()
history = []
```

7. Now we can proceed to the training loop of code, which should be relatively
    familiar by now.

```
for i in range(num_epochs):
    y_pred = net(x_train_t)
    loss = loss_fn(y_train_t,y_pred)
    history.append(loss.data)
    loss.backward()
    optimizer.step()
    optimizer.zero_grad()
```

```
test_loss = loss_fn(y_test_t,net(x_test_t))
if i > 0 and i % 100 == 0:
    print(f'Epoch {i}, loss = {loss:.3f}, test loss {test_loss:.3f}')
```

8. In the training block of code, we are again using the training set inputs (X) and labels (Y) to train the network. Along with this, we are also calculating `test_loss` by passing the test dataset in the network.

9. Running that last block of code will show you how the loss or error gets reduced from about 390 to around 1 or 2. This value represents the amount of value the house prediction was off. You should also notice that the test loss does not keep the same rate as the training loss. This is an example of our network overfitting.

From running this model, you can see our training dataset could be optimized too well. However, when we tested the test dataset against the same model, our results showed more error or loss. The reason for this is because the model is overfitting to the training data. Over- and underfitting of data is a problem we will chat about in the next section.

Over- and Underfitting

The reason we withhold a test set of data during training is to determine how well our model is fitting. If the model is fitting too well to the training data and then gives poor results with test data, that is called *overfitting*. Overfitting can be the result of a network memorizing the training dataset.

This memorization of data can happen when a network is too deep or large. While a large network can learn more quickly and adapt to larger data inputs, this comes with a price. That price is often in the form of overfitting or memorizing the data. For that reason, we almost always want to keep our networks as small as possible.

However, the inverse problem of overfitting is underfitting. This happens when a network is too small and unable to find optimum weights. Our initial XOR problem with a single perceptron was an excellent example of a network underfitting.

There is a balance to finding the optimum network size for the number of inputs for your problem. To find this balance, you will often need to use trial and error to grow and shrink your network in order to find the optimum size. We will look at other methods in later chapters that can help reduce over- and underfitting.

Try to go back to the previous exercise and adjust the size of the network from layers of 50 neurons to a smaller number like 30 or 20. This should reduce some of the overfitting to the training data, but then the opposite may happen, underfitting. See if you can balance out the number of neurons in each layer so that the training loss and test loss are closer.

Balancing a network's size can be a tricky endeavor and a problem we will look at continuously through this book. In the next section, we explore the form of regression called *logistic regression*, or regression with classes.

Classifying Classes

In many applications of data science such as machine learning and deep learning, we may want to output a class rather than a value. We saw in an earlier example how we may have a network that takes images of cats and dogs and then learns to classify them as cats or dogs. The process of doing this is called *classification* or regression with classes.

Regression with classes is like finding an equation, except in classification we use a form called *logistic regression*. Logistic regression is where we look for the equation that best separates the classes. We can then use that learned equation to divide input space into classes.

Figure 1-9 shows an example of a logistic regression function separating classes of dogs and cats. On one side of the equation or boundary we have cats and the other dogs. The classification of these images is done with a probability they reside in each class, and the measure we use to determine loss is called *cross-entropy loss*.

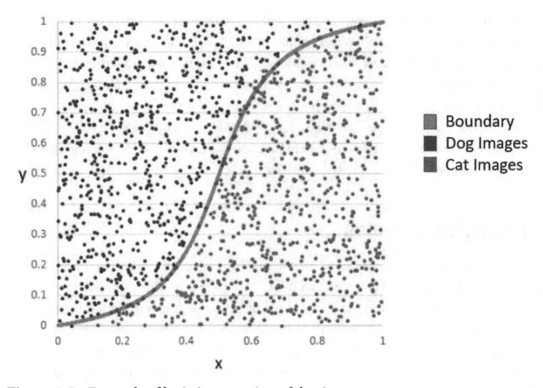

Figure 1-9. *Example of logistic regression of dog images*

To use logistic regression, the output of our network needs to be in the form of a probability or value from 0 to 1. That means when performing classification, the activation function in the last layer will often be sigmoid or a similar function. On top of this, we apply either a `BinaryCrossEntropy` for two classes or `CrossEntropy` loss for multiple classes.

For a single output, the following loss equation shows the function for determining a single class loss:

$$L = -y \cdot log\left(y_{pred}\right)$$

where:

y = the label value

y_{pred} = the predicted value

The dot symbol in the middle of the equation represents the inner product. To see how this works for multiple classes, we first need to understand how we encode classes in deep learning in the next section.

One-Hot Encoding

For the cross-entropy loss function to work, we need to encode our classes into a form called *one-hot encoding*. One-hot encoding is where we convert a class value to an array the same size as the number of classes, where the class value is represented as a 1 in that position and all other values are 0.

Figure 1-10 shows how we could one-hot encode a couple example digits from the MNIST dataset. For each image, the class is one-hot encoded where the 1 represents the position of the class.

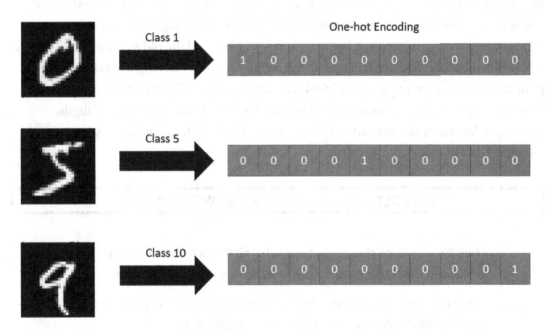

Figure 1-10. *One-hot encoding the MNIST digits*

In a classification scenario, we then reduce the output of the network to match the one-hot encoding class vectors. These vectors have the cross-entropy loss equation applied against them. As an example, consider a network that was fed an image from class 2, one-hot encoded as $[0,0,1,0,0,0,0,0,0,0]$. The output or y_pred value may end up being something like $[.2,.3,.6,.4,.1,.8,.1,.1,.1,.1]$. This in turn would predict a class of 6, because of the highest value of .8 being in the sixth position.

With cross-entropy loss, the loss for each value that is incorrect should be 0 and is accounted for as well as the loss from the missing classification. Thus, the error for each output neuron is accounted and the loss of each is pushed back through the network. We will see how this works in the next section.

Classifying MNIST Digits

In deep learning, we have several standard beginning datasets we can both learn from and test new methods with. As deep learning matures, these beginner or test datasets have gotten more sophisticated, but we can still use some of the old standards to learn by.

For the next exercise, we will look at the MNIST handwritten digits dataset to perform classification using a multiple-layer network. Each digit is a 28×28-pixel image that will represent 784 inputs into our network. Since there are 10 classes of digits, our output layer will also output to 10 neurons. We will see how this is constructed in Exercise 1-5.

EXERCISE 1-5. CLASSIFYING MNIST WITH PYTORCH

---

1. Open the GEN_1_classify_pytorch.ipynb notebook from the project's GitHub site. If you are unsure on how to access the source, check Appendix B.

2. In this example, we are going to use the torchvision.datasets module to load the MNIST dataset so the imports will be slightly different.

```
import os
import torch
import torch.nn as nn
from torch.autograd import Variable
import torchvision.datasets as dset
```

```
import torchvision.transforms as transforms
import torch.nn.functional as F
import torch.optim as optim
```

3. The next block of code will set up a data folder in the root and create it. This is where the next lines of code will download the dataset to. Before that, though, we use the transforms object to create a tensor transformation to transform the data. We need to do this because the data is represented as grayscale bytes with values of 0–255. The transform code normalizes this data to values between 0 and 1.

```
root = './data'
if not os.path.exists(root):
    os.mkdir(root)

trans = transforms.Compose([transforms.ToTensor(),
                            transforms.Normalize((0.5,), (1.0,))])
train_set = dset.MNIST(root=root,train=True,
                       transform=trans, download=True)
test_set = dset.MNIST(root=root,train=False,
                      transform=trans, download=True)

print(train_set)
print(test_set)
```

4. Moving down the next block creates the batches of data using the DataLoader class. DataLoader batches the data so that it can be more efficiently server to the network as well as provides options to shuffle the data. In this code block, we create two loaders, one for training and the other test.

```
batch_size = 128

train_loader = torch.utils.data.DataLoader(
                dataset=train_set,
                batch_size=batch_size,
                shuffle=True)
test_loader = torch.utils.data.DataLoader(
                dataset=test_set,
                batch_size=batch_size,
```

```
                        shuffle=False)
print(len(train_loader))
print(len(test_loader))
```

5. Our next step is to create a new MLP model. This network will take as inputs a digit of 28×28 pixels as input and push that down to 500 neurons and then from 500 to 256 and eventually down to 10. Remember the output of 10 neurons represents one output for each class in our dataset.

```python
class MLP(nn.Module):
  def __init__(self):
    super(MLP, self).__init__()
    self.fc1 = nn.Linear(28*28, 500)
    self.fc2 = nn.Linear(500, 256)
    self.fc3 = nn.Linear(256, 10)
  def forward(self, x):
    x = x.view(-1, 28*28)
    x = torch.relu(self.fc1(x))
    x = torch.relu(self.fc2(x))
    x = torch.sigmoid(self.fc3(x))
    return x

  def name(self):
      return "MLP"
```

6. With the MLP model class defined, we can move to instantiating the class as well as creating an optimizer and loss function. Notice the loss function is of type `CrossEntropyLoss` since this is a classification problem.

```python
model = MLP()
optimizer = optim.SGD(model.parameters(), lr=0.01, momentum=0.9)
loss_fn = nn.CrossEntropyLoss()
```

7. We next come to the training block of code, which again is quite similar to our last couple of exercises. In this example, since we are using a `DataLoader`, there are a few subtle differences you should note.

```python
epochs = 10
history=[]
for epoch in range(epochs):
```

```
avg_loss = 0
for batch_idx, (x, y) in enumerate(train_loader):
  optimizer.zero_grad()
  x, y = Variable(x), Variable(y)
  y_pred = model(x)
  loss = loss_fn(y_pred, y)
  avg_loss = avg_loss * 0.9 + loss.data * 0.1
  history.append(avg_loss)
  loss.backward()
  optimizer.step()
  if (batch_idx+1) % 100 == 0 or (batch_idx+1) == len(train_loader):
    print(f'epoch: {epoch}, batch index:' +
        f'{batch_idx+1}, train loss: {avg_loss:.6f}')
```

8. At this stage, add another code block to the notebook and use the code from the previous exercises to output a plot of the history.

9. As another extended exercise, you can also add code to test the test dataset to see how well your model is fitting to the data.

Aside from doing validation on the test set, we could also use the model to predict classes on known digits. This would allow us to see how the model was performing. We won't worry about going through those extra steps just yet, though. We will leave that for the heavy lifting and more thorough data visualizations in Chapter 2, where we start looking at the foundations of generative modeling.

Conclusion

Deep learning and neural networks have been brewing in the back of academia for decades, and it was not until relatively recently that their true power has been unleashed. Unleashing this power required the alignment of several factors such as improved GPU processes, big data, and data science as well as an increased interest in machine learning technologies. While these factors held up the science of deep learning, it was the discoveries in generative modeling that have really pushed the envelope and imagination of AI researchers.

In this chapter, we introduced deep learning and how it works, from the basic perceptron to the MLP and deep learning networks. We learned how to train these networks using supervised learning to perform regression and classification. For the rest of this book, we will advance to a form of learning called *unsupervised* and/or *self-supervised learning* as well as other advanced methods called *adversarial learning*.

In the next chapter, we explore the basics of unsupervised learning with autoencoders and then move on to adversarial learning with generative adversarial networks.

CHAPTER 2

Unleashing Generative Modeling

Deep learning was stuffed away in the back closet for years, often thought of as a pseudo or fuzzy science. It wasn't until the last decade that an explosion of growth has fostered in a new wave of AI. AI has matured from using supervised learning to more advanced forms of learning including unsupervised, self-supervised learning, adversarial learning, and reinforcement learning. It was from these other forms of learning that the area of generative modeling has come to flourish and advance in many areas.

The advent of deep learning and the innovative architectures it brings have allowed us to better understand what it means for an AI to learn. We previously struggled with implementing AI that could self-learn, but with deep learning, this can easily be accomplished. Furthermore, new and varied forms of learning have exploded. Adversarial learning, which has broad applications to generative modeling, was also born and in turn has fueled an AI explosion.

Generative modeling, by our most recent definition, is the building of an AI or machine learning model that can learn to generate new and varied content using a variety of learning methods. The key term here is *generate* and is not to be confused with filtering or transforming content. When we build a generative model, it needs to generate new content, sometimes completely randomly, but in many cases we conditionally control content generation.

In this chapter, we unleash generative modeling. We explore how to construct the base form of generators called *autoencoders*. Then we move on to learning how to better extract features from image data using convolution. From there we build an autoencoder with convolution layers added for better feature extraction. After that, we dive into the basis of adversarial learning and explore the generative adversarial network (GAN). We finish the chapter by upgrading a vanilla GAN to use deep convolution.

© Micheal Lanham 2021
M. Lanham, *Generating a New Reality*, https://doi.org/10.1007/978-1-4842-7092-9_2

As we introduce generative modeling in this chapter, we will look to explore the basis for other forms of learning from unsupervised to adversarial learning and GANs. These are the main the topics we will explore in this chapter:

- Unsupervised learning with autoencoders

- Extracting features with convolution

- The convolutional autoencoder

- Generative adversarial networks

- Deep convolutional GANs

Before continuing, be sure you have all the required knowledge of deep learning with PyTorch as covered in Chapter 1. In the next section, we begin our exploration into unsupervised learning and autoencoders.

Unsupervised Learning with Autoencoders

Supervised learning, or learning with labels, is a foundation to data science and a basis for many machine learning models. It is a natural form of learning and one that we often use ourselves to understand a problem space. When we or other animals do it, we call it *concept learning*. However, the concept of concept learning may blend with other forms of learning like unsupervised learning.

Unsupervised learning is a form of self or self-supervised learning that allows the machine to learn the concept of something by understanding the representation without labels. By not using labels, the machine is able to learn the representation of an object without the bias of a label. While in some cases we may use labels to control some aspect of that understanding, it is important to understand that labels are not used in the learning process.

To train a machine/model to learn the representation of something without a label, we build a machine that can decompose and then compose the model back together. The process for this in deep learning is referred to as *encoding* and *decoding* or *autoencoding*. An autoencoder is a machine that can encode or break down data into some lower form of representation. Then from that lower form it rebuilds it with a decoder.

Figure 2-1 shows a deep learning autoencoder. On the left side the input is a raw MNIST digit being fed into the encoder. The encoder decomposes the image into some latent or hidden form. From this the decoder decomposes the image back to the original as best it can, not by using labels but by understanding how well it can rebuild the image.

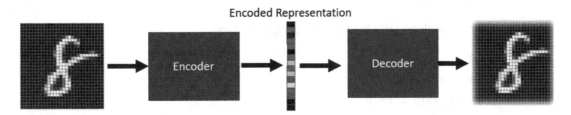

Figure 2-1. *Autoencoder architecture*

An autoencoder learns and improves on its model by how well it reconstructs the original image. In fact, we measure the error or loss in an autoencoder by doing pixel-wise comparisons between the original and recomposed images. With a deep learning autoencoder, we stack the two models of an encoder/decoder together.

For Exercise 2-1, we are going to use the MNIST fashion dataset with an autoencoder. This dataset is comprised of ten classes of garments you may wear or carry, from shoes, boots, and purses to sweaters, pants, and dresses. The images also provide excellent visuals of how an autoencoder can learn the recompositing process.

EXERCISE 2-1. A FASHION AUTOENCODER

---

1. Open the GEN_2_autoencoder.ipynb notebook from the project's GitHub site. If you are unsure on how to access the source, check Appendix B.

2. Run the first imports cell and then move to the next cell that sets up some image helper functions. The first function, imshow, is for rendering a PyTorch tensor image to a Matplotlib plot. The second function, to_img, converts the tensor to the appropriate size and dimensions.

    ```
    def imshow(img,size=10):
      img = img / 2 + 0.5
      npimg = img.numpy()
      plt.figure(figsize=(size, size))
    ```

```
    plt.imshow(np.transpose(npimg, (1, 2, 0)))
    plt.show()

def to_img(x):
    x = x.view(x.size(0), 1, 28, 28)
    return x
```

3. The next block contains the familiar hyperparameters:

```
epochs = 100
batch_size = 64
learning_rate = 1e-3
```

4. The next block of code downloads the data, transforms it, and batches it into a DataLoader. We then extract a batch of images from DataLoader and use the image util functions imshow and make_grid to render Figure 2-2. The make_grid function is part of the torchvision.utils module.

```
img_transform = transforms.Compose([
    transforms.ToTensor(),
    transforms.Normalize([0.5], [0.5])
])

dataset = mnist('./data', download=True, transform=img_transform)
dataloader = DataLoader(dataset, batch_size=batch_size, shuffle=True)

dataiter = iter(dataloader)
images, labels = dataiter.next()
imshow(make_grid(images, nrow=8))
```

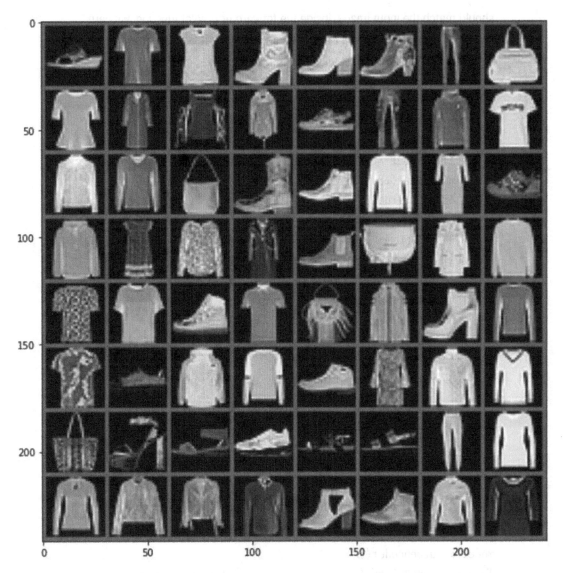

Figure 2-2. *Fashion MNIST dataset*

5. Next, we create a class for the autoencoder. Inside this class we construct two models, an encoder and decoder. The encoder reduces the input data from a vector of size 784 (28×28) down to 128 input neurons. This gets reduced further to 64, then to 12, and finally to an output vector of size 3. The decoder takes as input a vector of size 3, from the encoder. Then it builds it up to 12, 64, 128, and finally 784 outputs, where the complete output from the decoder

should match the input image. Inside the forward or predict function, we can see the encoder model being fed the image, encoding it, and then passing the output to the decoder that rebuilds the image.

```
class Autoencoder(nn.Module):
  def __init__(self):
    super(Autoencoder, self).__init__()
    self.encoder = nn.Sequential(
      nn.Linear(28 * 28, 128),
      nn.ReLU(True),
      nn.Linear(128, 64),
      nn.ReLU(True), nn.Linear(64, 12), nn.ReLU(True), nn.Linear(12, 3))
    self.decoder = nn.Sequential(
      nn.Linear(3, 12),
      nn.ReLU(True),
      nn.Linear(12, 64),
      nn.ReLU(True),
      nn.Linear(64, 128),
      nn.ReLU(True), nn.Linear(128, 28 * 28), nn.Tanh())

  def forward(self, x):
    x = self.encoder(x)
    x = self.decoder(x)
    return x
```

6. In the next cell, we instantiate the model and construct the loss function and optimizer.

```
model = Autoencoder()
loss_fn = nn.MSELoss()
optimizer = torch.optim.Adam(
    model.parameters(), lr=learning_rate, weight_decay=1e-5)
```

7. Now we can move on to the training code and where all the magic happens. Again, we see the familiar double loop, looping through epochs and then data. Inside that, we can see the data is extracted from the dataloader, but we ignore the labels. We wrap up the images and feed them to the model to output the

y_pred or decoded output. Then we take the MSE loss for each pixel in the image. Where the differences in pixel values are the loss, we train back to the network. The rest of the code resets the gradient and then pushes back the loss through the network.

```
for epoch in range(epochs):
  for data in dataloader:
    x_img, _ = data
    x_img = x_img.view(x_img.size(0), -1)
    x_img = Variable(x_img)
    # ===================forward=====================
    y_pred = model(x_img)
    loss = loss_fn(y_pred, x_img)
    # ===================backward====================
    optimizer.zero_grad()
    loss.backward()
    optimizer.step()
  # ===================log=========================
  clear_output()
  print(f'epoch [{epoch}/{epochs}], loss:{loss.data:.4f}')
  pic = to_img(y_pred.cpu().data)
  imshow(make_grid(pic))
```

8. As you run the training code cell, you will see how the model improves over time. Figure 2-3 shows the difference in the first training epoch to the last.

epoch [0/100], loss:0.1177

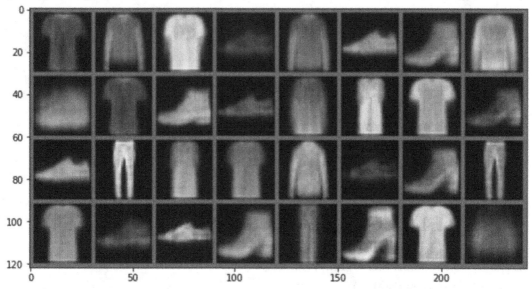

epoch [99/100], loss:0.0780

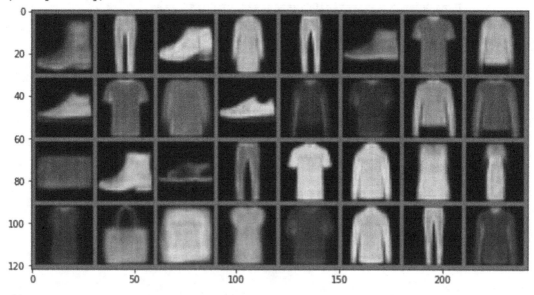

Figure 2-3. *Training the autoencoder*

At this stage, we have an autoencoder that can encode an image of some piece of clothing down to vector representation of size 3. That latent or hidden representation of the image may also be referred to as an *encoding* or *embedding*. We can use that learned encoding to represent characteristics of the images that make them unique, and the best part is the machine learns this on its own. In Chapter 3, we will look at more ways of understanding how these encodings can be visualized.

Our focus at this stage is to understand how an image can be decomposed into some lower representation and then rebuilt using unsupervised learning. That is, at no point did we label the images, and the model was learned entirely by understanding the reconstruction of images. This is an immensely powerful concept, and we use it throughout deep learning.

The process of learning to encode and decode data reduces dimensionality in data to understand how similar words are in natural language processing. It has also been extended to machine translation and text generation using transformers. Autoencoders are often the first demonstration of combining multiple forms of network architecture to solve some tasks.

While we were able to achieve some nice results with the networks we have used thus far, we need to discover new improvements to network layers that allow us to extract features from data. In the next section, we learn how using a convolutional filter can extract features in images and other data.

Extracting Features with Convolution

Up until 2012 image analysis with neural networks was done by flattening an image to a single one-dimensional (1D) vector. We have done this already several times with MNIST digits and fashion datasets. In those cases, we flattened the data from a 28×28-pixel image to a flat vector of 784 pixels. Figure 2-4 shows how an image is flattened in deep learning.

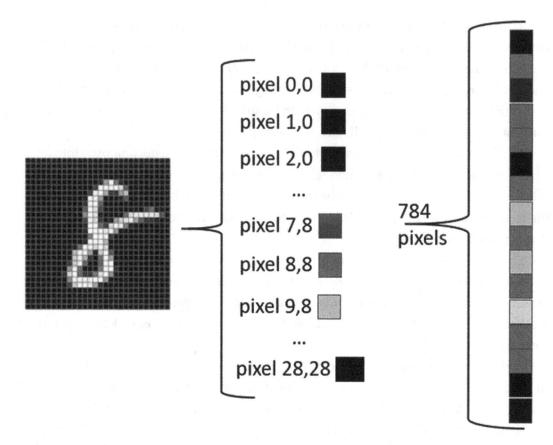

Figure 2-4. *Flattening an image to a 1D vector*

In Figure 2-4 we can see how the MNIST digit of 28×28 pixels is flattened into a single layer. This single layer of inputs is then fed into the neural network. From this our network can classify or rebuild the representation well, but not perfectly. In fact, the more complex the image, the more difficult this becomes. It was also a major limitation of deep learning since it often missed obvious image features.

That all changed in 2012 when a team led by Geoffrey Hinton won the ImageNet challenge by a wide margin. ImageNet is a labeled classification dataset of 1.5 million images in 1,000 classes. The challenge was being able to classify this dataset better than other models. Hinton's team used a new network type called a *convolutional neural network* (CNN).

CNN deep learning systems describe a type of network layer that can extract like or similar features from data using a process called *convolutional filtering*. In convolutional filtering, a patch or kernel of some dimension is passed over an image. As the kernel moves across the image, it uses a training set of weights to scale the values of the image as a type of filtering process.

Figure 2-5 shows the convolutional filtering process. In the figure we can see the kernel patch being slid across the image. As the filter is passed over the image, it multiplies the weights of the kernel with the pixel values of the image. This has the effect of filtering the image. The output of this filtering is shown in the next image. In this case, the filter may have been similar to an edge detection filter.

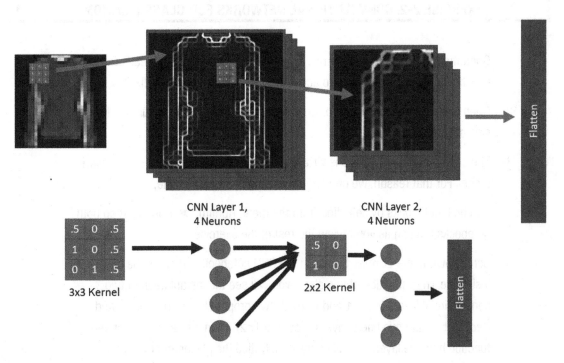

Figure 2-5. *Convolutional filtering process*

At the bottom of Figure 2-5, you can see how a single kernel filter with weights is like a single neuron/perceptron in a network layer, where each weight in the neuron is a value in the kernel. In the figure, we can also see how successive layers of convolution can be applied in sequence. This has the effect of filtering the extracted filtered features from the extracted.

This successive filtering process with convolutional layers has the effect of identifying key features in images, like a dog's ear or a cat's eyes. After all the features of the image or other data is extracted, the data is again flattened and fed into linear network layers.

For Exercise 2-2 we are going to revisit Exercise 1-5 and explore the difference using CNN layers will have. We are going to add a couple of two-dimensional (2D) convolutional layers for image extraction. From this, we can make comparisons against our previous exercise and get a sense for how much feature extraction improves performance.

EXERCISE 2-2. CONVOLUTIONAL NETWORKS FOR CLASSIFICATION

1. Open the GEN_2_classify_cnn.ipynb notebook from the project's GitHub site. If you are unsure on how to access the source, check Appendix B.

2. Open GEN_1_classify_pytorch.ipynb as well. We want to run a side-by-side comparison.

3. The code for both notebooks is virtually identical except for the neural network model. For that reason, we can omit reviewing most of the code.

4. Run both notebooks by selecting from the menu Runtime ➤ Run All. Keep both notebooks running as you review the rest of the exercise.

5. Scroll down to the ConvNet class in the CNN notebook and study the code. Inside the init function of this class, you will see the instantiation of two convolutional layers (conv1 and conv2), two dropout layer (dropout1 and dropout2), and two linear layers (fc1 and fc2). Then we see in the forward function how the layers are all connected. Notice the placement of the dropout layers and the linear layers as well.

```
class ConvNet(nn.Module):
  def __init__(self):
    super(ConvNet, self).__init__()
    self.conv1 = nn.Conv2d(1, 32, 3, 1)
    self.conv2 = nn.Conv2d(32, 64, 3, 1)
    self.dropout1 = nn.Dropout2d(0.25)
```

```
        self.dropout2 = nn.Dropout2d(0.5)
        self.fc1 = nn.Linear(9216, 128)
        self.fc2 = nn.Linear(128, 10)

    def forward(self, x):
      x = self.conv1(x)
      x = F.relu(x)
      x = self.conv2(x)
      x = F.relu(x)
      x = F.max_pool2d(x, 2)
      x = self.dropout1(x)
      x = torch.flatten(x, 1)
      x = self.fc1(x)
      x = F.relu(x)
      x = self.dropout2(x)
      x = self.fc2(x)
      output = F.log_softmax(x, dim=1)
      return output
```

6. Notice how the inputs are defined into a Conv2D layer. Figure 2-6 shows how
    each input into Conv2D is configured and defined for the layer. CNN layers
    consume channels of data, where typically a black and white or grayscale
    image like MNIST has only one channel. For color images we will break the
    image into red, green, and blue color channels with their respective color
    values. We will explore more about channels later.

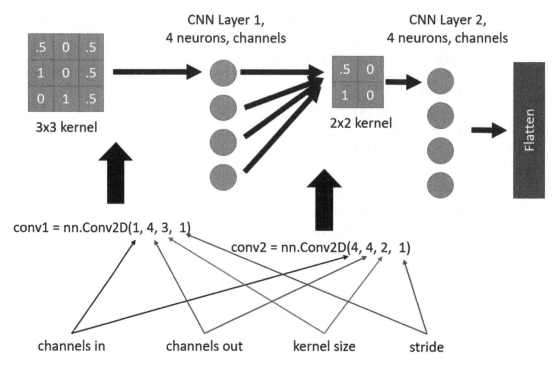

Figure 2-6. *PyTorch CNN layer configuration*

7. Look back at the ConvNet class code, and notice how the input is a single
    channel that extracts 32 output channels. This is pushed into a second Conv2D
    layer with 32 input channels and 64 output channels. The output of this last
    layer of convolution creates 9,216 inputs when flattened into the first linear
    layer. Then the second layer consumes those with 128 neurons and outputs 10
    classes.

Let both examples run to completion, and you will notice two things. The first is that
our previous exercise runs much quicker, and the second is the results with convolution
are ten times better. Convolution requires more parameters and thus more training
time. These training times can be dramatically reduced by using CUDA or a GPU. In later
sections and chapters, we will use GPU processing where applicable.

In the previous exercise, we also used a new layer called Dropout. Dropout layers are
special layers that randomly turn off a percentage of neurons for each training iteration.
The number of random neurons turned off is set at the instantiation and is typically .2 to

.5 or 20 to 50 percent. By turning off random neurons in each training pass, the network has to shift to becoming more generalized and less specialized. Dropout layers are the first cure we select if we diagnose a network as overfitting or memorizing data points. (In Chapter 1 we discussed over- and underfitting.)

Figure 2-7 shows a visualization from the TensorSpace.js Playground at `https://tensorspace.org/html/playground`. This is an excellent resource to look at and understand how convolution and deep learning layers do what they do.

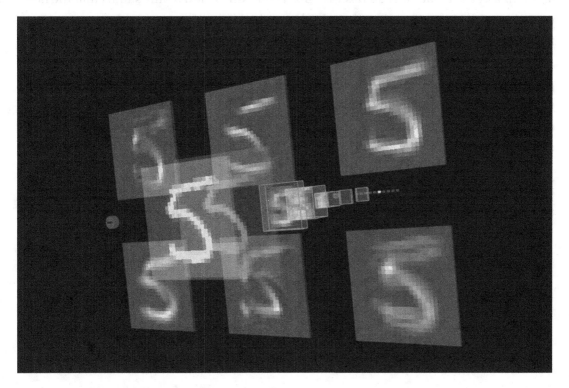

Figure 2-7. *TensorSpace.js Playground*

Convolutional layers can help us classify but also can provide added benefit when doing generative modeling. However, when we generate content with convolution, we must apply the opposite of filtering called *transposition*. We will cover using CNN in autoencoders with transposition in the next section.

The Convolutional Autoencoder

While convolution can certainly help us identify features in images, it can also help us do the opposite: generate features. In our previous autoencoder exercise, you can clearly see how the images are built up on pixel-by-pixel representations.

Review Figure 2-3 again, and you can see a fuzzy outline around the images or features; this is characteristic for the type of generation we applied. After all, all we did in our first autoencoder was do pixel-by-pixel comparisons so it only stands that the best we can hope for is an averaging of pixel values, which in turn causes the fuzzy areas in the images.

For us to extract and then generate features, we will look at upgrading the plain-vanilla autoencoder we used last time to work with convolution. As a further demonstration, we will also upgrade our base dataset to CIFAR10. The CIFAR dataset consists of 28×28 color images separated into ten classes, from airplanes to dogs and cats.

Figure 2-8 shows an example of the CIFAR10 dataset images with respective labels and was generated from our next exercise. In Exercise 2-3 we upgrade the vanilla autoencoder to use convolution to extract and learn how to generate features.

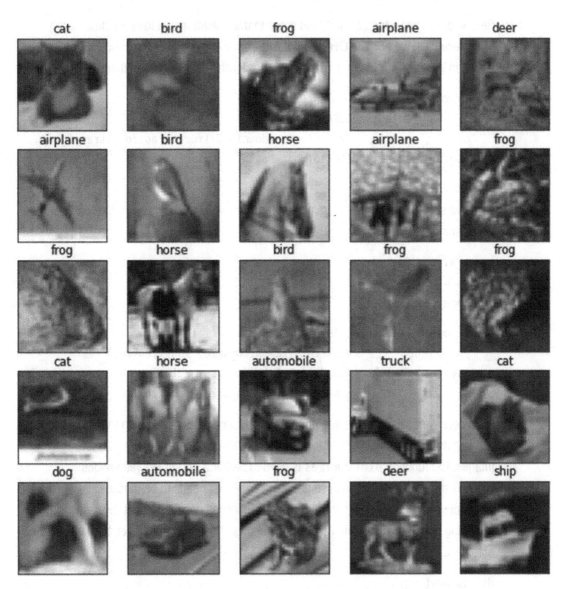

Figure 2-8. *Example images from the CIFAR10 dataset*

EXERCISE 2-3. CONVOLUTIONAL AUTOENCODER ON CIFAR10

---

1. Open the GEN_2_conv_autoencoder.ipynb notebook from the project's
    GitHub site. From the menu, select Runtime ➤ Run All to run the entire sheet. If
    you are unsure on how to access the source, check Appendix B.

2. Review the import's top cell and then move to the cell where we load the data.
    In this exercise, we pull the CIFAR10 dataset from the `torchvision` module.
    Notice in this case we just use a default tensor for the data transformations.

```
transform = transforms.ToTensor()
train_data = datasets.CIFAR10(root='data', train=True,
                              download=True, transform=transform)
test_data = datasets.CIFAR10(root='data', train=False,
                              download=True, transform=transform)
```

3. Next, we set the hyperparameters and instantiate the `DataLoaders` for the test
    and training sets.

```
batch_size = 64
epochs = 100
learning_rate = 1e-3

train_loader = torch.utils.data.DataLoader(train_data,
                        batch_size=batch_size, shuffle=True)
test_loader = torch.utils.data.DataLoader(test_data,
                        batch_size=batch_size, shuffle=True)
```

4. From here we again create some image display functions for rendering the
    image sets. We also provide text for the labels so that we can label the dataset
    images. The function `plot_images` is used to render a grid of labeled images.
    At the bottom of this function is some timing code that needs to be in place
    for rendering in the notebook during training. This code does not affect the
    rendering but just pauses the training code enough to output the images to the
    notebook.

```
def imshow(img):
    plt.imshow(np.transpose(img, (1, 2, 0)))  # convert from Tensor image

# specify the image classes
classes = ['airplane', 'automobile', 'bird', 'cat', 'deer',
           'dog', 'frog', 'horse', 'ship', 'truck']

def plot_images(images, labels, no):
  rows = int(math.sqrt(no))
  plt.ion()
```

```
fig = plt.figure(figsize=(rows*2, rows*2))
for idx in np.arange(no):
    ax = fig.add_subplot(rows, no/rows, idx+1, xticks=[], yticks=[])
    imshow(images[idx])
    ax.set_title(classes[labels[idx]])
time.sleep(0.1)
plt.pause(0.0001)
```

5. With the helper functions created, we can generate Figure 2-8 with the following code:

```
dataiter = iter(train_loader)
images, labels = dataiter.next()
images = images.numpy() # convert images to numpy for display

plot_images(images,labels,25)
```

6. That takes us to the ConvAutoencoder class, which is an upgrade from our earlier Autoencoder. In this class, we can see the linear layers have been swapped out for Conv2d layers. We can also see the addition of a new layer type called MaxPool2d or a pooling layer. Pooling layers are used to collect like features and are an efficiency in network training. With the decoder model, we can see the use of ConvTranspose2d layers. Convolution transpose layers are the opposite of convolution, and instead of extracting features, they generate learned features. In the decoder, you can see two ConvTranspose2d layers that output to a sigmoid activation function.

```
class ConvAutoencoder(nn.Module):
  def __init__(self):
    super(ConvAutoencoder, self).__init__()
    self.encoder = nn.Sequential(
      nn.Conv2d(3, 16, 3, padding=1),
      nn.ReLU(True),
      nn.MaxPool2d(2, 2),
      nn.Conv2d(16, 4, 3, padding=1),
      nn.ReLU(True),
      nn.MaxPool2d(2, 2)
    )
```

```
        self.decoder = nn.Sequential(
          nn.ConvTranspose2d(4, 16, 2, stride=2),
          nn.ReLU(True),
          nn.ConvTranspose2d(16, 3, 2, stride=2),
          nn.Sigmoid()
        )

    def forward(self, x):
      x = self.encoder(x)
      x = self.decoder(x)
      return x

model = ConvAutoencoder()
print(model)
```

7. Before training, we need to create our loss function and optimizer with the following code:

```
loss_fn = nn.MSELoss()
optimizer = torch.optim.Adam(model.parameters(), lr=learning_rate)
```

8. Finally, we can move to the last cell, `training`. This code is the same as our previous autoencoder exercise, but we only show it here for review:

```
for epoch in range(epochs):
  train_loss = 0.0
  for data in train_loader:
    images, labels = data
    optimizer.zero_grad()
    generated = model(images)
    loss = loss_fn(generated, images)
    loss.backward()
    optimizer.step()
    train_loss += loss.item()*images.size(0)

  train_loss = train_loss/len(train_loader)
  clear_output()
  print(f'Epoch: {epoch+1} Training Loss: {train_loss:.3f}')
  plot_images(generated.detach(),labels,16)
```

As the code is running, you will see output like Figure 2-9. Figure 2-9 shows the training transition for `ConvAutoencoder`. Notice how the images now look more blocky than fuzzy with the images being more a collection of recognizable features than fuzzy outlines. Because of the size of the input images, we are somewhat limited to how well we can apply convolution in this exercise. In later chapters and exercises, we will see various configurations of CNN layers.

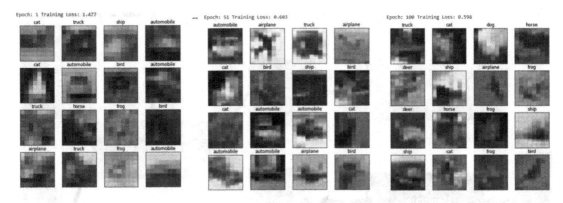

Figure 2-9. *Convolutional autoencoder training on CIFAR10*

You can see that we could approach limits to the detail we use with convolution, which has more to do with the localized feature extraction. Localized feature extraction limits the layer to extracting features in blocks. In later chapters, we will look at other methods that can alleviate issues like these.

Autoencoders use a form of learning called *unsupervised* since the model does not depend on image labels to rebuild or regenerate the images. However, the model still needs an image as a sample to encode and then decode from. What we want to explore next is how we can generate images from a random latent space by using adversarial learning.

Generative Adversarial Networks

Generative adversarial networks have been around for only a short time. GANs are similar to autoencoders in that they comprise two models, but instead of an encoder and decoder, they use a discriminator and generator.

Figure 2-10 shows the architecture of a GAN. An analogy we often use to help describe the GAN process is that of the interaction between an art verifier and the art forger. In this scenario, the art forger works to create a convincing fake piece of art the verifier will consider to be real. The art verifier learns to understand what is real by looking at real art. In turn, the forger learns by creating pieces of art and then testing them against the verifier.

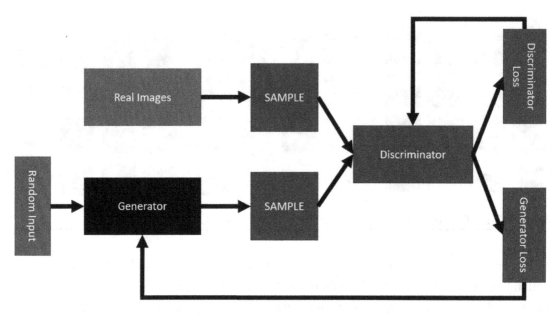

Figure 2-10. *GAN architecture*

In a GAN, the art forger is the generator. The generator generates images or artwork from a random latent space we define as Z. After an image is generated, it is shown to the verifier or discriminator. The discriminator learns by looking at samples of real images as well as the fake images generated by the discriminator.

The discriminator determines loss by analyzing the real and fake images. If it thinks a real image is fake, it outputs higher loss back to itself. Likewise, if the generator passes a fake image as real, it returns a higher loss back to the discriminator but a lower loss to itself, the generator. Both models learn in tandem to get better at either generating images or determining what images are real or fake.

While the architecture of the GAN may seem overly complex, it is just the autoencoder architecture split apart and reversed, where the generator is the same as the decoder and the discriminator resembles the encoder. Thus, the discriminator is like the encoder, but instead of outputting an encoding, it just outputs real or fake. On the other side, the generator consumes a random latent space of data, resembling nothing, and learns to create new images that can be passed off as real to the discriminator.

In Exercise 2-4, we are going to build a vanilla or base GAN. With this base form of the GAN, we will learn how to generate digits from the MNIST handwritten dataset. We won't construct a GAN class but instead create two classes for the generator and discriminator and then write code to train them independently as well.

EXERCISE 2-4. BUILDING A VANILLA GAN

1. Open the GEN_2_vanilla_gan.ipynb notebook from the project's GitHub site. From the menu, select Runtime ➤ Run All to run the entire sheet.

2. Skipping the imports and the imshow helper function at the top of the sheet, we come to the data loading and transformation code. All of this code should be review at this point.

```
transform = transforms.Compose([
                transforms.ToTensor(),
                transforms.Normalize((0.5,),(0.5,))
                ])
to_image = transforms.ToPILImage()
trainset = MNIST(root='./data/', train=True, download=True,
transform=transform)
train_loader = DataLoader(trainset, batch_size=100, shuffle=True)

device = 'cuda'
```

3. Next, we come to the Generator class. The generator serves the same roles as the decoder in an autoencoder. The key difference with a generator is the input is just random noise. It may help to think of this random noise as a random thought vector. Other than that, the code is quite similar to the decoder we have seen before. The size of this random vector is defined by latent_dim in the inputs. You may also notice the use of a new activation function LeakyReLU. This function allows for some amount of negative leakage

through a function rather than cutting values off at 0. For the forward/prediction function, you can see the outputs are reshaped to a 1,28,28 tensor of the image.

```python
class Generator(nn.Module):
  def __init__(self, latent_dim=128, output_dim=784):
    super(Generator, self).__init__()
    self.latent_dim = latent_dim
    self.output_dim = output_dim
    self.generator = nn.Sequential(
        nn.Linear(self.latent_dim, 256),
        nn.LeakyReLU(0.2),
        nn.Linear(256, 512),
        nn.LeakyReLU(0.2),
        nn.Linear(512, 1024),
        nn.LeakyReLU(0.2),
        nn.Linear(1024, self.output_dim),
        nn.Tanh()
    )

  def forward(self, x):
    x = self.generator(x)
    x = x.view(-1, 1, 28, 28)
    return x
```

4. After that comes the discriminator, which resembles the encoder from an autoencoder, with the difference being the output is only two classes, real or fake. In the forward/predict function, the input is reshaped from a 28×28 image to an input vector of size 784. Other than that, the architecture of the model resembles a typical classifier, which is what it also is. In fact, a fully trained discriminator can have excellent applications in classifying data as real or fake for other applications outside of GANs, perhaps being able to determine if some input image is of a particular type based on the data fed the GAN. You could therefore train a discriminator on faces, and it would learn to recognize a real or fake face.

```python
class Discriminator(nn.Module):
  def __init__(self, input_dim=784, output_dim=1):
    super(Discriminator, self).__init__()
```

```python
        self.input_dim = input_dim
        self.output_dim = output_dim
        self.discriminator = nn.Sequential(
          nn.Linear(self.input_dim, 1024),
          nn.LeakyReLU(0.2),
          nn.Dropout(0.3),
          nn.Linear(1024, 512),
          nn.LeakyReLU(0.2),
          nn.Dropout(0.3),
          nn.Linear(512, 256),
          nn.LeakyReLU(0.2),
          nn.Dropout(0.3),
          nn.Linear(256, self.output_dim),
          nn.Sigmoid()
        )

    def forward(self, x):
        x = x.view(-1, 784)
        x = self.discriminator(x)
        return x
```

5. With the classes built, we can move on to instantiate them this time with GPU (cuda) support for more performant training. In addition, we will create the optimizer and loss function.

```python
generator = Generator()
discriminator = Discriminator()

generator.to(device)
discriminator.to(device)

g_optim = optim.Adam(generator.parameters(), lr=2e-4)
d_optim = optim.Adam(discriminator.parameters(), lr=2e-4)

g_losses = []
d_losses = []

loss_fn = nn.BCELoss()
```

6. We also need to create some helper functions. The first function `noise` generates the random noise we input into the generator. The second function `make_ones` is a helper to mark the batch as real, with 1s. Then the third function, `make_zeros`, does the opposite and marks the batch of images with a zero for fake.

```python
def noise(n, n_features=128):
    return Variable(torch.randn(n, n_features)).to(device)

def make_ones(size):
    data = Variable(torch.ones(size, 1))
    return data.to(device)

def make_zeros(size):
    data = Variable(torch.zeros(size, 1))
    return data.to(device)
```

7. Then we add an extra helper function to train the discriminator. Recall that the discriminator is trained on both the real images as well as the fake images from the generator. Then notice that we pass the real data into the discriminator and use that to predict the real loss, `loss_real`. That loss is then backpropagated through the discriminator. After that, we test a set of fake images and pass that loss backward to the network. Then we return the combined output of both losses. You should also notice the use of the functions `make_ones` and `make_zeros` being used to label the data as real (1) or fake (0).

```python
def train_discriminator(optimizer, real_data, fake_data):
    n = real_data.size(0)
    optimizer.zero_grad()

    prediction_real = discriminator(real_data)
    loss_real = loss_fn(prediction_real, make_ones(n))
    loss_real.backward()

    prediction_fake = discriminator(fake_data)
    loss_fake = loss_fn(prediction_fake, make_zeros(n))

    loss_fake.backward()
    optimizer.step()

    return loss_real + loss_fake
```

8. We also create a helper function to train the generator. In the second train function, we only need to pass the fake/generated images into the discriminator to evaluate the loss. Notice the use of the helper function make_ones being used to label the data as real.

```
def train_generator(optimizer, fake_data):
    n = fake_data.size(0)
    optimizer.zero_grad()

    prediction = discriminator(fake_data)
    loss = loss_fn(prediction, make_ones(n))

    loss.backward()
    optimizer.step()

    return loss
```

9. Finally, we come to the training code, which is slightly different than we've seen before. Since we constructed two helper functions to independently train the generator and discriminator, this code loops through and calls those *functions*. Notice the addition of an internal training loop restricted by k. This internal loop can be used to increase the number of iterations we train the discriminator on each epoch. We may want or need to do this to better balance the training. It is best to keep both models in a GAN learning at the same rate.

```
epochs = 250
k = 1
test_noise = noise(64)

generator.train()
discriminator.train()
for epoch in range(epochs):
    g_loss = 0.0
    d_loss = 0.0
    for i, data in enumerate(train_loader):
        imgs, _ = data
        n = len(imgs)
        for j in range(k):
            fake_data = generator(noise(n)).detach()
            real_data = imgs.to(device)
```

```
            d_loss += train_discriminator(d_optim, real_data, fake_
            data)
        fake_data = generator(noise(n))
        g_loss += train_generator(g_optim, fake_data)

    img = generator(test_noise).cpu().detach()
    g_losses.append(g_loss/i)
    d_losses.append(d_loss/i)
    clear_output()
    print(f'Epoch {epoch+1}: g_loss: {g_loss/i:.8f} d_loss: {d_loss/
    i:.8f}')
    imshow(make_grid(img))
```

Figure 2-11 shows the output generation at the start of training, then to epoch 150 or so, and at the final 250 epochs. As you run the exercise, you will see the output update over time as well. Notice how the images start out very rough and random but over training become quite readable as handwritten digits.

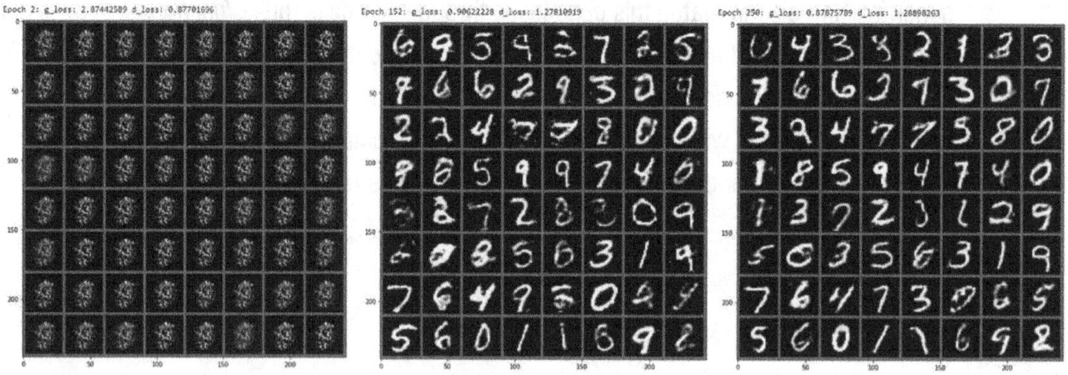

Figure 2-11. *Training a GAN on MNIST*

Another key difference to understand with a GAN is that the images are entirely generated from random noise. Remember the input into our generator is just random noise. Over time the generator has learned to convert that noise to realistic-looking digits. Keep in mind those digits were not drawn but were generated from essentially nothing.

Once you get the basics of a GAN, you can start to understand its potential to generate anything. Indeed, the number of GAN applications and variants are exploding daily. In the next section, we look at a first-step improvement on the vanilla GAN.

Deep Convolutional GAN

Deep convolutional GANs (DCGANs) are a first-level improvement to the vanilla GAN by adding convolutional layers. Just like our previous work on autoencoders when we added convolution, the architecture changes are almost exact. That means the encoder/discriminator will convolve and extract features, while the decoder/generator will transpose to build features.

The DCGAN is virtually identical to other features in training, with the one exception being we may alter the input of random noise size. We can use the same base of code and upgrade it to a DCGAN in our next exercise. With this upgrade we will also want to increase the size of the input images. Bigger images allow for more applications of convolution and thus feature extraction.

For Exercise 2-5 we have three options for the real set of training data to the discriminator. Our ideal dataset will be the CelebA dataset, which is a collection of celebrity faces. However, CelebA is a busy dataset and not always accessible to download. Therefore, we also provide a couple of other options such as CIFAR10 and STL10. The STL10 dataset is the same as CIFAR, but the images are larger.

EXERCISE 2-5. GENERATING FACES WITH DCGAN

---

1. Open the GEN_2_DCGAN.ipynb notebook from the project's GitHub site. From the menu, select Runtime ➤ Run All to run the entire sheet.

2. The imports are almost identical to the previous exercise with one key difference. We abstract the dataset import as DS so that we can easily swap data sources from CelebA, CIFAR10, or STL10. If you want to use a different dataset, change the import dataset.

   ```
   from torchvision.datasets import CelebA as DS  #other options
   CIFAR10, STL10
   ```

3. The next major change is the transform we apply to the input dataset. For the import of this data, we transform the data to an `image_size` of 64 for this version. Then we use `CenterCrop` to crop the image and finally normalize it.

```
transform=transforms.Compose([
                            transforms.Resize(image_size),
                            transforms.CenterCrop(image_size),
                            transforms.ToTensor(),
                            transforms.Normalize((0.5, 0.5, 0.5),
                            (0.5, 0.5, 0.5)),
                ])
```

4. Figure 2-12 shows an excerpt from the output of the data loading and then the output of the real images.

Figure 2-12. *CelebA sample faces*

5. Next, we come to the updated Generator class. We can see the inputs are updated to include a new input called feature_maps. The input feature_maps sets the number of channels we pass between convolutional layers. You can treat this number as a hyperparameter for tuning on other datasets. There is also the addition of a new layer type called BatchNorm2d. This new layer renormalizes the data as it passes through and is a way for us to limit the loss gradient getting too big or too small.

As a network gets deeper with more layers, there becomes a higher possibility that the loss gradient either gets too small or gets too big. This is called *vanishing* or *exploding* gradients. By normalizing the data as it passes through the network, we avoid the vanishing/exploding gradients by keeping the weight parameters closer to zero. An added benefit of this is increased training performance.

```
class Generator(nn.Module):
  def __init__(self, latent_dim=100, feature_maps=64, channels=3):
    super(Generator, self).__init__()
    self.main = nn.Sequential(
    nn.ConvTranspose2d( latent_dim,
                        feature_maps * 8, 4, 1, 0, bias=False),
    nn.BatchNorm2d(feature_maps * 8),
    nn.ReLU(True),
    nn.ConvTranspose2d(feature_maps * 8,
                        feature_maps * 4, 4, 2, 1, bias=False),
    nn.BatchNorm2d(feature_maps * 4),
    nn.ReLU(True),
    nn.ConvTranspose2d( feature_maps * 4,
                        feature_maps * 2, 4, 2, 1, bias=False),
    nn.BatchNorm2d(feature_maps * 2),
    nn.ReLU(True),
    nn.ConvTranspose2d( feature_maps * 2,
                        feature_maps, 4, 2, 1, bias=False),
    nn.BatchNorm2d(feature_maps),
    nn.ReLU(True),
    nn.ConvTranspose2d( feature_maps, channels, 4, 2, 1, bias=False),
    nn.Tanh()
  )

  def forward(self, input):
    return self.main(input)
```

65

6. From the generator we can move on to the updated discriminator. For the most part, this will resemble the classifier and encoder we built in earlier exercises. Notice that the discriminator is deeper and has more layers in this example, and we can do that since the size of the base image is 64×64. The larger the base image into the network, the more convolutional layers we can use to extract features.

```python
class Discriminator(nn.Module):
  def __init__(self, feature_maps=64, channels=3):
    super(Discriminator, self).__init__()
    self.main = nn.Sequential(
    nn.Conv2d(channels,
              feature_maps, 4, 2, 1, bias=False),
    nn.LeakyReLU(0.2, inplace=True),
    nn.Conv2d(feature_maps,
              feature_maps * 2, 4, 2, 1, bias=False),
    nn.BatchNorm2d(feature_maps * 2),
    nn.LeakyReLU(0.2, inplace=True),
    nn.Conv2d(feature_maps * 2,
              feature_maps * 4, 4, 2, 1, bias=False),
    nn.BatchNorm2d(feature_maps * 4),
    nn.LeakyReLU(0.2, inplace=True),
    nn.Conv2d(feature_maps * 4,
              feature_maps * 8, 4, 2, 1, bias=False),
    nn.BatchNorm2d(feature_maps * 8),
    nn.LeakyReLU(0.2, inplace=True),
    nn.Conv2d(feature_maps * 8, 1, 4, 1, 0, bias=False),
    nn.Sigmoid()
    )

  def forward(self, input):
    return self.main(input)
```

7. The rest of the code is almost the same as the code for the vanilla GAN with one subtle difference. In the DCGAN, we construct the input noise using a slightly different method, as shown here:

```python
noise = torch.randn(n, latent_dim, 1, 1, device=device)
```

8. Figure 2-13 shows the output of the DCGAN as it is training on faces. You may also notice in the training code how we limit the number of batches or samplings with a new hyperparameter called num_sumples. This hyperparameter controls how many sample batches to pull from the train_loader. The more samples, the better the training. However, more samples also means much slower training. Therefore, you may want to tune this feature for best results.

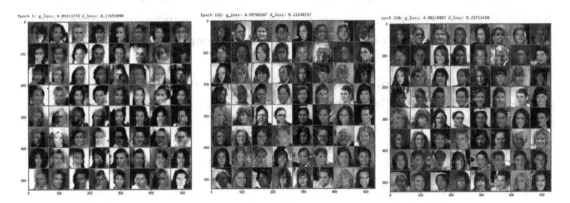

Figure 2-13. *Training progression of DCGAN*

9. Feel free to go back and alter the DS dataset to either CIFAR10 or STL10 to see different results. Are the results better or worse than you expected? Would they fool you?

It may take the DCGAN several thousands of epochs to get 100 percent convincing images. At the 250-epoch stage, you should be able to pick a couple generated images that do resemble real pictures of faces. Be sure to note how the DCGAN trains, and look at the differences in output.

In addition to the suggested three datasets, you could try any other sources of real images to train the DCGAN on. All you need to do is make sure the DataLoader can load the image data you need for training, thus allowing you to train the DCGAN on any other three channel image sources.

The DCGAN is the first variation in GANs that we will look at. It is a simple architecture change that allows us to better process and generate images. Other variation of GANs we will study in future chapters may come with architectural and method differences improving on the GAN.

Conclusion

In this chapter, we looked at first-order generative modeling with autoencoders and GANs. We learned how we can adapt supervised learning methods to use unsupervised methods like a GAN. While an encoder encodes content to some latent hidden space, the GAN creates content from a nothing/random hidden space. This allows for all manner of content generation using all manner of real base images.

Ideally, we would like to control what a GAN generates by controlling the random input in some manner. If we could learn to control the random thoughts the generator used to output certain types of images, we could control generating those types of images.

In the next chapter, we will look at controlling the hidden input space of the generator to alter the output by attributes. This will give us the ability to control if a female or male face is generated with glasses or without.

CHAPTER 3

Exploring the Latent Space

It is easy to get caught up in the mystique of artificial intelligence and machine learning when you view the results of an autoencoder or generative adversarial network. The results of these deep learning systems can appear magical and even show signs of actual intelligence. Unfortunately, the truth is far different and even challenges our perception of intelligence.

As our knowledge of AI advances, we often set the bar higher as to what systems demonstrate intelligence. Deep learning is one of those systems that many now view as a mathematical process (machine learning) rather than actual AI. In this book, we won't debate if DL is AI or just machine learning. Instead, we will focus on why understanding the math of DL is important.

To build generators that work, we need to understand that deep learning systems are nothing more than really good function solvers. While that may dissolve some of their mystique, it will allow us to understand how to manipulate generators to do our bidding.

In this chapter, we take a closer look at how deep learning systems learn and indeed what they are learning. We then advance to understanding variability and how DL can learn the variability of data. From there we use that knowledge to build a variational autoencoder (VAE). Then we learn how to explore the variability to tune the VAE generator. Extending that knowledge, we move on to building and understanding a conditional GAN (CGAN). We finish the chapter using the CGAN to generate pictures of food.

We have a lot of ground to cover in this chapter, from some math theory to more complex examples. Here is a quick overview of the main topics we will cover in this chapter:

- Understanding what deep learning learns
- Building a variational autoencoder

© Micheal Lanham 2021
M. Lanham, *Generating a New Reality*, https://doi.org/10.1007/978-1-4842-7092-9_3

- Learning distributions with the VAE

- Variability and exploring the latent space

It is recommended that you have reviewed the material in Chapters 1–2 before reading this chapter. You may also find it helpful to review the math concepts of probability statistics and calculus. Don't feel you need to be proficient in calculus; just understand the basic concepts. In the next section, we take a closer look at the math of deep learning.

Understanding What Deep Learning Learns

The math of deep learning is by no means trivial and in many ways was a major hurdle the technology struggled with for decades. It wasn't until the development of automatic differentiation that deep learning really evolved. Before that, people would spend hours tuning the math for just the simplest of networks. Even the simplest of networks we learned in previous chapters could have taken days or weeks to get the math right.

Now, with automatic differentiation and frameworks like PyTorch or TensorFlow, we can build powerful networks in minutes. What is more, we can teach and empower people with deep learning far quicker. You no longer need a PhD to train networks, and in fact it is common for kids in elementary school to employ deep learning experiments.

Unfortunately, with the good also comes the bad. Automatic differentiation allows us to treat a deep learning system as a black box. That means we may lack some subtle understanding for how a network learns, thereby potentially missing obvious problems in our system or worse yet encounter long struggles trying to fix issues we lack the understanding to resolve.

Fortunately, understanding the math of how a deep learning system learns can be greatly simplified. While it may be difficult to explain these concepts to a five-year-old, anyone who can read a graph and a simple equation will likely understand the concepts in the next section.

Deep Learning Function Approximation

Critical to understanding what deep learning does is understanding that a network is just a good function approximator. In fact, simply put, that is all deep learning systems do. They just approximate known or unknown functions well. Now this can be a bit confusing when we think about how neural networks classify, so let's look at an example.

Consider Figure 3-1, which classifies images of cats and dogs into two groups using a function called *logistic regression*. In the figure we can see the boundary line representing the function output. On one side of this function the images of cats will fall and on the other the images of dogs. This would be an example of a deep learning network that learns an unknown function. We know there is a function that can define the classification between cats and dogs visually and mathematically; we just don't know what it is.

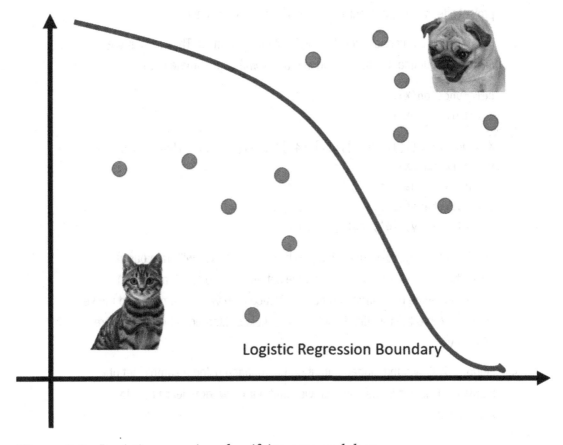

Logistic Regression Boundary

Figure 3-1. *Logistic regression classifying cats and dogs*

When a network learns to classify images, it learns the logistics or set of logistic regression functions that break things into classes. While we know there is a set of functions that define the boundary to those classes, in most cases we don't worry about the exact function. Instead, through training and learning, our deep learning system learns or approximates the function on its own.

To demonstrate how networks do function approximation, we are going to jump into another code example. Exercise 3-1 borrows from a previous regression example, and instead of using data, we can define a known function. Using a known function will demonstrate with certainty how our networks learn.

EXERCISE 3-1. DEEP LEARNING FUNCTION APPROXIMATION

1. Open the GEN_3_function_approx.ipynb notebook from the GitHub project site. If you are unsure how, then consult Appendix B.

2. Run the whole notebook by selecting Runtime ➤ Run all. Then look at the top cells, which define a simple function that we will approximate here:

```
def function(X):
  return X * X + 5.

X = np.array([[1.],[2.],[3.],[4.],[5.],[6.],[7.],[8.],[9.],[10.]])
y = function(X)
inputs = X.shape[1]
y = y.reshape(-1, 1)
plt.plot(X, y, 'o', color='black')
```

3. The function function defines a parabolic equation we will train our network on. In the next cell, you can see how we set up hard-coded X inputs 1–10 that we use to define our learned outputs, y. Notice how we can simply feed the set of inputs X into the function to output our labels or learned values, which are labeled y.

4. Figure 3-2 shows the plotted output of the equation using a sample set of inputs (1–10). This is the equation we want our network to learn how to approximate.

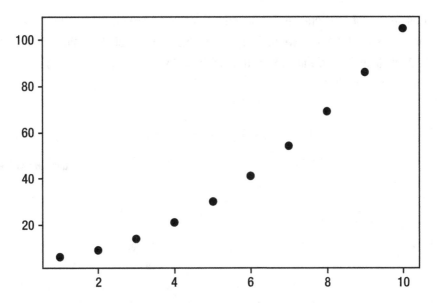

Figure 3-2. *Plotted output of equation values*

5. We have already reviewed most of the remaining code in previous chapters. Our focus now will be on identifying slight differences. Notice how we still break our input data into training and testing splits with the following:

```
X_train, X_test, y_train, y_test = train_test_split
(X, y, test_size=0.2, random_state=0)
num_train = X_train.shape[0]
X_train[:2], y_train[:2]
num_train
```

6. With only 10 input points, it may seem like a silly step to break the data up into 8/2 points, respectively. However, even when tuning to a known function, it can help validate our results, so we still like to do validation/test splits.

7. The only other change to this notebook is another step of validation we do at the end with the following code:

```
X_a = torch.rand(25,1).clone() * 9
y_a = net(X_a)
y_a = y_a.detach().numpy()
plt.plot(X_a, y_a, 'o', color='black')
```

8. This section of code creates 25 random values using `torch.rand`. Since the values are from 0–1, we multiply the output by 9 to scale the values from 0–9. We then run the values X_a through the network `net` and plot the output shown in Figure 3-3.

```
[<matplotlib.lines.Line2D at 0x7f13ced56cc0>]
```

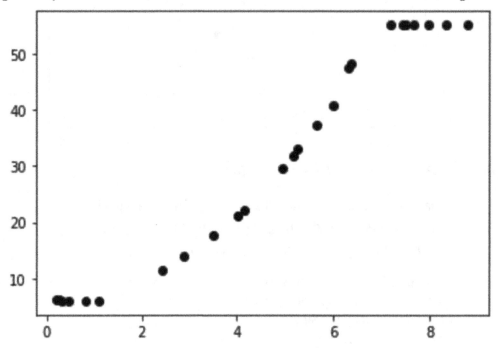

Figure 3-3. *Plotted output of test results on trained network*

9. Notice in Figure 3-3 how the network is excellent at approximating the middle values of our known function. However, you can see that the network struggles at the bottom (close to X of 0) and the top end. You can see that the network has greater difficulty approximating the boundaries. Yet it still does very well matching the function's middle values.

What we just demonstrated is how a simple network can learn to approximate a function. However, we did notice how the network still struggled to approximate the boundaries, and we need to understand why that is.

Part of the problem is the limited amount of data we used to train the network. After all, we used only 10 points. However, the function approximation through the center is very good compared to what happens at the ends. So, why is that?

Simply put, a deep learning system can approximate well to a known or unknown equation within the bounds of the known data or limits of the equation. In our last example, we used data values from 0–10, with bounds of 0 and 10. Yet we see from Figure 3-3 that the function approximation falls down around an input of X=7 or so. As well, on the lower bounds we can see an issue when X<1.

The reason for these inconsistencies is not about the data or indeed the limits to the data but rather the calculus itself. Remember our discussion about auto-differentiation with calculus in Chapter 2? Now the problem isn't the auto-part of differentiation but rather the differentiation itself.

Calculus helps us to understand the rate of change in an equation or function. This is useful for everything from launching spaceships to building better buildings and of course deep learning. It is also important for us to understand why calculus is used to approximate the functions in deep learning, which is covered in the next section.

The Limitations of Calculus

If you have studied calculus at all, you have learned the limitations of calculus. One of them is that all functions we differentiate with calculus need to be continuous. Figure 3-4 shows example functions with areas called *discontinuities*. These areas are examples of functions that would prevent a deep learning system from learning or being able to approximate the function.

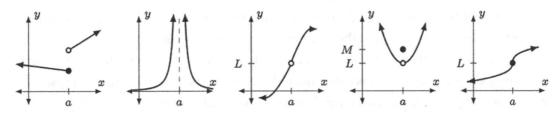

Figure 3-4. Examples of discontinuous functions

In generative modeling, you can think of a discontinuity as a gap or vacuum in the data. Now don't confuse a gap in this context with space or distance. We use the term *gap* here to refer to a break where nothing in the function has any sample values.

The term *space* or *latent space* or *hidden space* is used to define the unknown area in a continuous set of data.

Generally, when we see black regions in generated output, that represents a problem with limits or discontinuities in the input data.

Understanding the difference between continuous and discontinuous data will take some time, and sometimes it may not be so obvious. The key takeaway here is to understand that function approximation has limits, and those limits are often defined by the data itself, sometimes requiring a network to approximate a continuous function over discontinuous data.

Now, getting back to limits, let's understand how limits can affect how a network learns or approximates to data. In our exercise example, our network had limits of 1 and 10. If we refer to Figure 3-3, we can see that our function flattens at X<1, which makes sense considering the limit of 1. This means our network just limits the function approximation to y=5 when X is less than 0.

What is not so obvious, however, is why our function approximation fails on values greater than 7 when the limit was 10. To understand why this issue arises, we need to understand how deep learning uses calculus to approximate functions, which is covered in the next section.

Deep Learning Hill Climbing

At the heart of any deep mathematical concept there is always a grounded intuition that explains the why. Unfortunately, most classic mathematics courses often miss explaining the why. Instead, they rely on the student to uncover the intuition on their own through performing proofs or regurgitating equations. This is something that works well for budding mathematicians but tends to overwhelm everyone else. For that reason, we will always focus more on the why of the math rather than the specific how.

Calculus is a tool we use in deep learning systems to balance the weights/parameters of a network. Recall that calculus works by understanding the rate of change of a system. In deep learning, calculus allows us to determine how much change and to where we need to adjust the weights in a network.

Calculus and gradient optimization or gradient descent is not the only tool we can use to find the weights in a network. It is currently the most efficient, but deep learning systems have and continue to use other machine learning methods. If you are interested in learning more, Google *deep learning PSO* or *deep learning GA*.

Figure 3-4 shows a one-dimensional function with a ball at position 1. We can think of the goal of a deep learning system as guiding that ball over the function. Further yet, we want the ball to be able to follow that function as close as possible. The better that ball is able to follow the function path, the better our networks will predict or generate content.

Notice that in Figure 3-4 we are moving the ball an amount calculated through automatic differentiation and our optimization method multiplied by the learning rate (alpha). This is labeled "Gradient X alpha" in the figure. The amount of gradient can be altered by selecting different optimization methods and/or tuning the hyperparameters. Tuning the optimizer, be it stochastic gradient descent or Adam, is something we will cover later.

That leaves us with *alpha*, the learning rate or learning parameter. This hyperparameter allows us to control how much tuning a network uses to map that ball to a function. In Figure 3-5 and in our previous example, we are applying a high learning rate. As you can see, this is demonstrated with the movement of the ball to position 2 and then 3. Notice how the ball moves well above the function line in the third position. This is what causes our problem in the previous example.

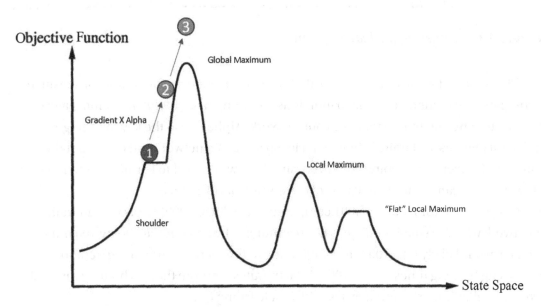

Figure 3-5. *Deep learning hill climbing*

In Figure 3-6, we can see how reducing the learning rate allows the network to better approximate the function path. You can see with the addition of the boundaries it also makes it critical that our ball stays with bounds. When the ball bounces outside the maximum bounds, the value becomes maxed. This is the reason for the maximum values not mapping well in Figure 3-3.

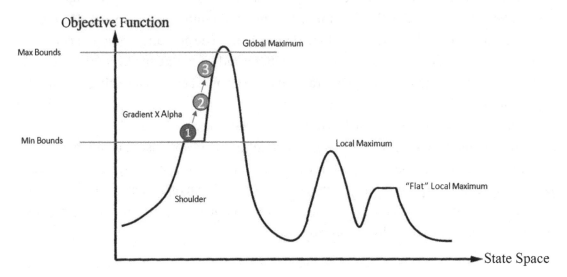

Figure 3-6. *Adjusting the learning rate*

The primary takeaway from this is that the learning rate is often our most sensitive hyperparameter when mapping to functions. Learning rate, or *alpha*, is almost always the first to tune and understand for your network. Alpha is also the key to tuning how quickly a network will train. When alpha is too large, the network will overshoot the function. Conversely, keeping alpha very small allows the ball to learn the function well. However, this can come at great cost of computing performance.

Consider a learning rate of .001 compared to a value of .00001. Say we can train a network with the smaller rate in, say, 10 minutes. The smaller rate, being 100 times smaller, would likely take 100 times longer due to the increase in training iterations, increasing training times to $10 \times 100 = 1000$ minutes, or more than 16 hours. This is all from a smaller but unfortunately less efficient learning rate.

To demonstrate this key concept further, let's look at another example. Exercise 3-2 shows how we will tune the alpha (learning rate) to better approximate the function in our previous exercise.

EXERCISE 3-2. TUNING ALPHA FOR EFFICIENT LEARNING

1. Open the GEN_3_learning_rate.ipynb notebook from the GitHub project site. If you are unsure how, then consult Appendix B.

2. Run the whole notebook by selecting Runtime ➤ Run all. Notice that where we define the optimizer, as shown in the following code, is where we set the learning rate lr.

```
loss_fn = nn.MSELoss()
optimizer = torch.optim.Adam(net.parameters(), lr = .001)
```

3. In our previous example of GEN_3_function_approx.ipynb, our learning rate was .025, a value that was on the cusp of being able to correctly map the function.

4. Our goal here is to find the learning rate that will allow the network to map 100 percent to the function or close to it.

5. Change the learning rate lr in the previous block of code and rerun the entire notebook by selecting Runtime ➤ Run all.

6. See if you can tune the lr value to match or approximate Figure 3-7. What happens if you make the learning rate really small?

7. If your network completes training before finding the function, this is easily fixed by increasing the number of epochs. Change num_epochs to something larger and see what results this has.

```
num_epochs = 8000
y_train_t = torch.from_numpy(y_train).clone().reshape(-1, 1)
```

8. Attempt to find the smallest number of epochs and the greatest learning rate that will train the network to match the function x2 + 5.

9. If you are feeling more advanced, try altering the number of neurons in each layer or altering the network itself.

```
[<matplotlib.lines.Line2D at 0x7efcecd63278>]
```

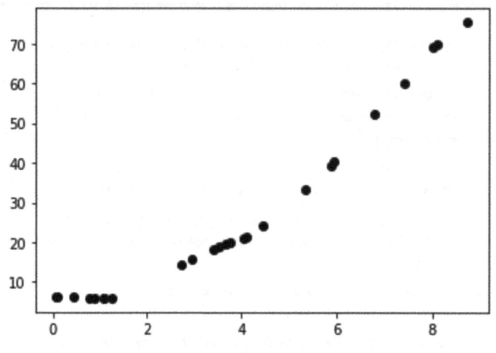

Figure 3-7. *Better approximation to the function x2 + 5?*

As you worked through the previous exercise, it may have occurred to you that there may be limitations to the network itself and the data. In fact, that is exactly the case in the exercise. For this particular example, the network attempts to overfit the data. We will look at how to resolve both over- and underfitting in the next section.

Over- and Underfitting

Oftentimes when newcomers design networks, their first assumption is that having more neurons means better results. Unfortunately, this is rarely the case for a number of reasons that we will understand in later chapters. For now, we want to focus on how the network size itself can overfit or underfit to the function.

You may often hear deep learners mention their network is over- or underfitting to the data. What they really mean is that the network is over- or underfitting to the function that matches or categorizes that data.

Figure 3-8 shows our previous example plotted against the function x2 + 5 to see how well we were actually matching to the results. As you can see, we are not matching well—in fact not really well at all considering the size of our network. However, the size of our network isn't the entire problem. Network size and the amount of data go almost hand in hand.

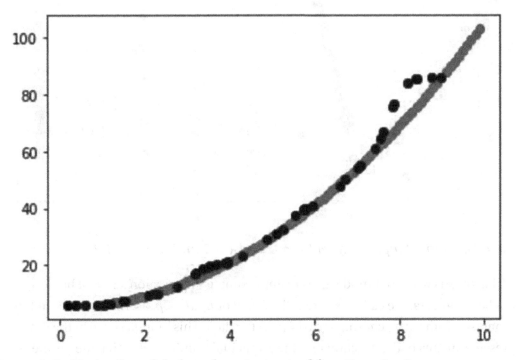

Figure 3-8. *Trained model plotted against actual function x2 + 5*

Think back to our previous exercise and our attempts to approximate to the function x2 + 5. Our first couple of attempts used a set number of data points from 1 to 10. We fed those data points into the function to generate y outputs. This generated our set of trainable data, which we then broke up into training and testing splits.

Figure 3-9 shows the plot of the 10 data points we generated to train and approximate the function. Of those points, we are removing 20 percent or 2 for test. This decreases the number of points in the dataset and in essence the size of space in between our training data.

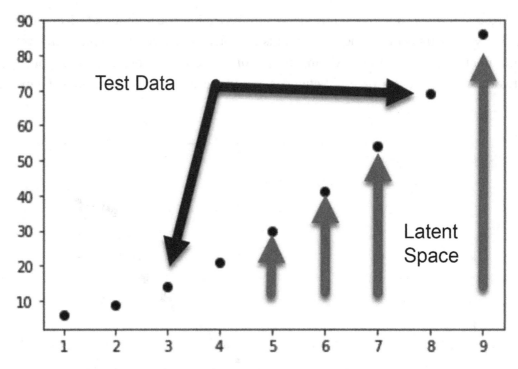

Figure 3-9. *Plot of input dataset from Exercise 3-1 and Exercise 3-2*

That means when our model goes to approximate the function x2 + 5, it has large areas between data. We call these areas the *latent* or *hidden space*. Again, this is not to be confused with a gap, chasm, or vacuum in the data. This hidden space in the data represents an area of the function missing any actual training data. That means the network must approximate values between these data points.

As it turns out, a big part of generative modeling is understanding how to extract results from this latent space, as we will see throughout this book. It's not unlike how we tried to extract the results of our previous function approximation exercise. Our previous assumption may have been to reduce the number of training points, but as we can see now, this increases the size of the latent space, causing our model to overfit to the function.

Overfitting is caused when a model approximates to a function but wrongly fills or approximates the latent space. In our previous function approximation examples, this was clearly the case, which was caused by two problems. The first was the lack of data, which caused the network to estimate large areas of latent space. The second was the size of our network, which had too much power to estimate the space in data. So, it essentially made up values, which can be a problem.

When a network is large, it has a lot of power, in this case too much power. Large networks can memorize data of all forms including images or video. This may be useful for networks scoped to specific data, but in generative modeling as well as most other machine learning disciplines, our goal is always to generalize.

Generalizing is always a goal we will undertake when building models. That means the data we feed into the model needs to be well distributed. Well-distributed data provides us with even latent space between data. In our previous example, our data was distributed into 10 bins, with values from 1–10. However, as we can see, our data was not distributed well enough.

Generalizing also means that our network should be only the size it needs, no more or less. If our network is too large, it will tend to overfit or fill in those latent space with incorrect values. Likewise, networks that are too small will underfit. Underfitting is often a consequence of too large of gaps or latent space in a model.

Figure 3-10 shows the classic example of underfitting to the x2 + 5 function by using a linear model or approximating a line. Underfitting in this example would be the result of the model not being complex enough. However, it can also occur when a network is not being fed enough data or that data is poorly distributed.

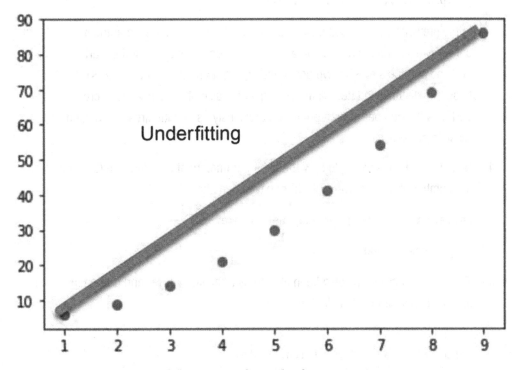

Figure 3-10. *A linear model trying to fit to the function x2 + 5*

To understand this in a tangible manner, our next exercise will focus on how we create models that over- and underfit to a set of function data. We will continue to use the base of our previous exercise, but this time we will modify some details to understand the problem better; see Exercise 3-3.

EXERCISE 3-3. UNDERSTANDING OVER- AND UNDERFITTING

1. Open the GEN_3_over_under.ipynb notebook from the GitHub project site. If you are unsure how, then consult Appendix B.

2. Run the whole notebook by selecting Runtime ➤ Run all. Notice how we changed the inputs (X,y) in the block just before the output plot, as shown here:

```
data_step = .1
X = np.reshape(np.arange(1,10, data_step), (-1, 1))
y = function(X)
inputs = X.shape[1]
plt.plot(X, y, 'o', color='red')
```

3. In the previous block of code, we alter the NumPy calls so that we autogenerate our array using np.arange. The arange function takes as input the start and end and the step size. We use a new hyperparameter called data_step to set the distance in latent space. Using a step size of .1 allows us to create 10 times the number of data points we previously used, thus greatly reducing those latent spaces.

4. Next, notice in this example how the hyperparameters of learning rate (lr) and the number of epochs (num_epochs) have been altered:

```
optimizer = torch.optim.Adam(net.parameters(), lr = .01)

num_epochs = 1000
```

5. The network design has also been altered by removing a layer and putting in another hyperparameter called neurons.

```
neurons = 20
torch.set_default_dtype(torch.float64)
net = nn.Sequential(
```

```
    nn.Linear(inputs, neurons, bias = True), nn.ReLU(),
    nn.Linear(neurons, neurons, bias = True), nn.Sigmoid(),
    nn.Linear(neurons, 1)
)
```

6. The neurons hyperparameter allows us to quickly alter the network size, allowing us to reduce the number of neurons to either over- or underfit the network.

7. Try adjusting just the two hyperparameters, data_step and neurons, to see what effect this has on over- or underfitting the model. Keep in mind that larger and complex models generally require more training iterations, num_epochs. This in turn means you may want to alter the learning rate lr, as well.

8. Now try to find the smallest network, the least number of neurons, that can learn the function the quickest, with fewer num_epochs. Feel free to adjust the learning rate (lr) and data_step as you need.

9. Finally, set data_step to a high value like 1.0 and then see what effect altering the neurons has. Do you need to increase or decrease the number of neurons?

The intention of the previous exercise was to demonstrate how our networks can over- and underfit the function that maps to our data. In this simple example, we could see quick and explicit feedback for what each of our hyperparameters did, allowing us to understand how a network can over- or underfit.

As we progress through this book and use more complex forms of data, like images, over- and underfitting often becomes less obvious. Fortunately, there are several clues that can help us identify these types of problems that we explore as we progress to those exercises.

One key element we didn't cover and can cause a lot of over/underfitting issues is data distribution. In the examples, the data was uniformly distributed across a range of values. This is rarely something we see in the real world.

For example, assume you had a dataset of 30,000 animal images (10,000 cats, 15,000 of dogs, and 5,000 birds) that you wanted to classify into cats, dogs, and birds. The problem is your data is not evenly distributed. You have more pictures of dogs, and hence your model will overestimate or memorize dogs. Instead, what we should do is reduce the number of cat and dog images to 5,000 to match the number of bird images. You could still argue that the cat and dog are more visually similar than a bird, thus potentially influencing your results.

We use and understand how data is distributed in several ways in generative modeling. It is in fact a foundation of many of the methods. In the next section, we will understand why data distributions are important to the variational autoencoder.

Building a Variational Autoencoder

In our previous example exercises, we explored how networks perform function approximation and discovered the latent space at discrete intervals. We found that by increasing our amount of data sampling intervals we could reduce our unknown latent space and improve our model training. This is a technique that works well for simplistic functions but does not scale to more complex problems.

A better approach to understanding our problem and the data itself is with statistics. Statistics allows us to understand the model and latent space in the data. In turn, statistics can provide us with a summary or overview of the models and the way we generate new data.

Statistics are fundamental to data science, deep learning, and generative modeling. In data science we use statistics to make decisions on what data to use. For deep learning, statistics are used to measure a model's efficiency, while generative modeling uses statistics to understand what the model is learning.

Now before we jump into the statistics of generative modeling and understanding latent spaces, it would be helpful to look at a working example. This example also can take a while to train, so while it does, you can read up on the theory in the next section.

In our next exercise, we are going to dive in and build a variational autoencoder (VAE). A VAE is like an autoencoder in structure and function except in how it learns that encoding part. Remember in an AE, the model learns how to encode a latent representation of the data before decoding it back to the original. In a VAE, the model learns how the latent encoding is distributed. We won't worry about understanding the distributed part just yet but instead jump in and see how the code works in Exercise 3-4.

EXERCISE 3-4: BUILDING A CONVOLUTIONAL VARIATIONAL AUTOENCODER

1. Open the GEN_3_conv_VAE.ipynb notebook from the GitHub project site.

2. Run the whole notebook by selecting Runtime ➤ Run all. Just below the first import cell, you will notice a cell checking for the device type, as shown in the following code:

```
device = torch.device('cuda' if torch.cuda.is_available() else 'cpu')
device
```

3. For this notebook and all future notebooks, we will be configuring the runtime to use a GPU. Figure 3-11 shows the runtime type accessible from the menu by selecting Runtime ➤ Change runtime type.

Figure 3-11. *Changing the runtime type on a Colab notebook*

4. Since we covered the next few sections of data handling code before, we will jump down to the cell that defines the ConvVAE class and the __init__ function.

```
class ConvVAE(nn.Module):
  def __init__(self, image_channels=3, h_dim=1024, z_dim=32):
    super(ConvVAE, self).__init__()
    self.encoder = nn.Sequential(
      nn.Conv2d(image_channels, 32, kernel_size=4, stride=2),
      nn.ReLU(),
      nn.Conv2d(32, 64, kernel_size=4, stride=2),
      nn.ReLU(),
      nn.Conv2d(64, 128, kernel_size=4, stride=2),
      nn.ReLU(),
      nn.Conv2d(128, 256, kernel_size=4, stride=2),
      nn.ReLU(),
      Flatten()
    )

    self.fc1 = nn.Linear(h_dim, z_dim)
    self.fc2 = nn.Linear(h_dim, z_dim)
    self.fc3 = nn.Linear(z_dim, h_dim)

    self.decoder = nn.Sequential(
      UnFlatten(),
      nn.ConvTranspose2d(h_dim, 128, kernel_size=5, stride=2),
      nn.ReLU(),
      nn.ConvTranspose2d(128, 64, kernel_size=5, stride=2),
      nn.ReLU(),
      nn.ConvTranspose2d(64, 32, kernel_size=6, stride=2),
      nn.ReLU(),
      nn.ConvTranspose2d(32, image_channels, kernel_size=6, stride=2),
      nn.Sigmoid(),
    )
```

5. This code is like the convolutional autoencoder we looked at before. The key difference here is the use of the three nn.Linear layers that now are responsible for learning the middle encoding distribution. Previously we defined a discrete data size as our learning representation. Have some patience: we will talk more about learning distributions later.

6. Jump over the remaining functions in the ConvVAE class and down to the block of code where the optimizer is defined.

```
optimizer = torch.optim.Adam(model.parameters(), lr=learning_rate)

def loss_fn(recon_x, x, mu, logvar):
  BCE = F.binary_cross_entropy(recon_x, x, size_average=False)
  KLD = -0.5 * torch.mean(1 + logvar - mu.pow(2) - logvar.exp())
  return BCE + KLD, BCE, KLD
```

7. We can see in the previous code the definition of a specialize loss function
    `loss_fn`. The code in this function is responsible for learning latent encoding
    distribution. Again, details to come later.

8. Finally, we come to the last block of the training code. This section is almost
    identical to our previous exercises. One key difference to note, however, is the
    highlighted line starting with `images =`. This line converts the images from a
    CPU tensor to a GPU tensor, the difference being where the memory resides and
    how it is processed. Tensors processed on GPUs can increase performance by
    more than 100x.

```
for epoch in range(epochs):
    train_loss = 0.0
    for data in train_loader:
        images, labels = data
        optimizer.zero_grad()
        images = images.to(device)
        generated, mu, logvar = model(images)
        loss, bce, kld = loss_fn(generated, images, mu, logvar)
        loss.backward()
        optimizer.step()
        train_loss += loss.item()*images.size(0)

    train_loss = train_loss/len(train_loader)
    clear_output()
    print(f'Epoch: {epoch+1} Training Loss: {train_loss:.3f}')
    plot_images(generated.cpu().data,labels,16)
```

9. Keep the notebook training as you continue reading through the next section.
    Be sure to check back periodically to see how the model progression learns. If
    you are not happy with the results, keep the model training by running the last
    cell. Each run of the last cell will add another 100 iterations of training to the
    model.

You can keep incrementally training your model if the notebook's runtime is not reset or expires. The runtime can be manually reset in the notebook or Google will expire it for you over time. Google suggests a runtime is good for 12 hours, but several factors appear to affect that. A few can be the number of notebooks you are running, if they are using GPUs and traffic.

Figure 3-12 shows the result of training the model after 100, 200, and 500 epochs. Notice how the VAE learns over iterations and try to pick out similarities and differences in the output. In fact, it almost looks like our model is going to the optometrist and being fitted for different levels of glasses. Each training iteration is fitting a better pair of glasses and thus making the images less fuzzy and more distinguishable.

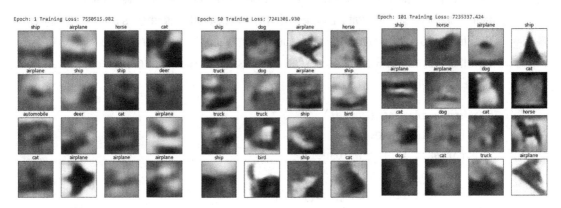

Figure 3-12. *Training results of VAE after 1, 50, and 101 iterations*

Now that we have seen how a VAE can function, we want to understand how the internal model works and learn the distribution or variation of the data in the next section.

Learning Distributions with the VAE

In a VAE, the model learns by not just understanding a middle or latent space encoding but rather how the encoding itself is distributed. By understanding the distribution or variability of the encoding, we can then learn to map across the latent encoding and generate those hidden spaces.

Figure 3-13 shows an AE and VAE side-by-side. In the figure, x represents the input, and x' represents the generated output. Likewise, z represents the latent encoding or middle representation of our data. In a VAE, there is a slight difference where we don't learn the encoding but rather learn the distribution or variability of that encoding. Then we use those learned parameters to sample a z representation and feed it into the decoder.

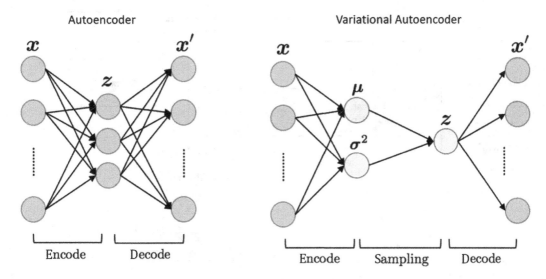

Figure 3-13.   *Comparison of AE vs. VAE*

The parameters this VAE model learns are the mean (μ) and standard deviation or variance squared (σ^2). If you recall from statistics, these are the same parameters we use to define a normal distribution. We often default to the normal distribution, but it is important to understand that data can be distributed in a myriad of ways.

Figure 3-14 shows an example of various distributions along with the normal or Gaussian distribution. The normal distribution is what we typically assume most data to resemble. Any use of basic descriptive statistics will generally assume data to be normally distributed.

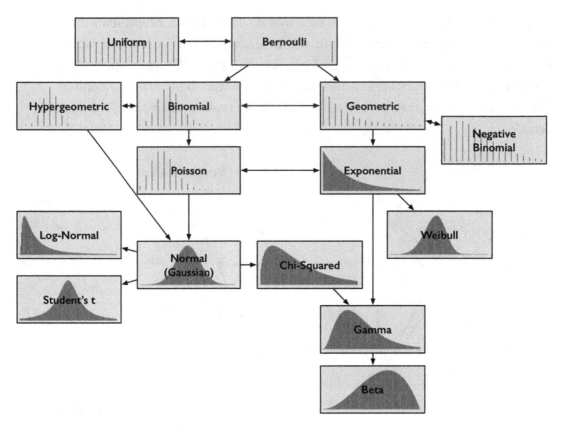

Figure 3-14. *Examples of various distributions*

A normal distribution as shown in Figure 3-15 is defined by two parameters, the mean (μ) and standard deviation (σ2). In a VAE, we let the network learn what the mean and standard deviations are of the distribution by using those values to generate a sample encoding. That sample encoding is then fed into the decoder, and a generated image is output.

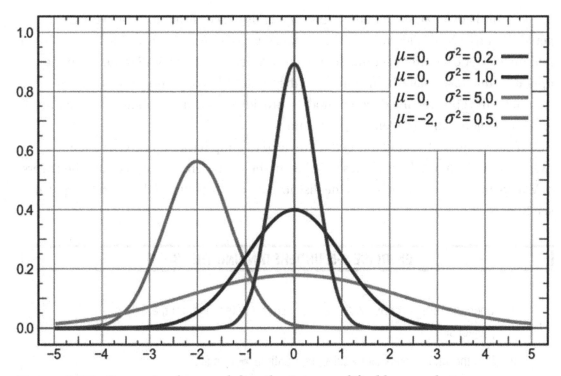

Figure 3-15. *Example of normal distributions modified by μ and σ2*

In Figure 3-15 we can see how different variations of μ and σ2 alter the shape of the normal distribution, where the mean always represents the center of the distribution and the standard deviation represents the spread or diversity.

Typically, we use statistics to find how data is distributed or what variability it represents. With a VAE we are still learning what the parameters of the data are, except now the data space we are focusing on becomes the hidden or latent space representation of that data. This has three advantages over trying to learn the entire latent space of a set of training images.

First, by learning the embedding space, the amount of data or dimensions of data can be greatly reduced. This allows our VAE to learn how features are represented in a simpler form, which in turn allows us to model a simple normal distribution. More complex forms of data could require a VAE to model more complex distributions or sets of distributions.

The second advantage provides us with the ability to map or generate from this learned latent space into new output. All we need to do is understand and control those distributional parameters, and we have new novel outputs.

The third advantage (but perhaps less of one) is the confirmation that we can generate output by just learning the input data distribution and converting it into normal parameters. One key architectural difference between an AE and VAE is the sampling operation that takes place in the middle of the model. This operation essentially decouples the encoder and decoder models, providing us with the ability for better reuse and further generative modeling opportunities.

Of course, to understand this better, we need to jump back into another code exercise and see how all this comes together. In the next exercise, we go under the covers of VAE to understand how it learns the distribution or variability of the encoded space; see Exercise 3-5.

EXERCISE 3-5. UNDERSTANDING THE VAE

1. Open the GEN_3_conv_VAE_latent.ipynb notebook from the GitHub project site.

2. Run the whole notebook by selecting Runtime ➤ Run all.

3. Jump down to the class definition code block and the encode and decode functions shown here:

```
def bottleneck(self, h):
    mu, logvar = self.fc1(h), self.fc2(h)
    z = self.reparameterize(mu, logvar)
    return z, mu, logvar

def encode(self, x):
    h = self.encoder(x)
    z, mu, logvar = self.bottleneck(h)
    return z, mu, logvar

def decode(self, z):
    z = self.fc3(z)
    z = self.decoder(z)
    return z
```

```
def forward(self, x):
    z, mu, logvar = self.encode(x)
    z = self.decode(z)
    return z, mu, logvar
```

4. Notice how the network restricts using the `bottleneck` function to generate the `mu` and `logvar` or variance parameter. Those parameters are then used to sample a z or encoded representation from the distribution. Then in the forward function we can see how encode is used to generate those parameters as well as the sampled z, with z being fed into the decoder as the encoding used to generate new output.

5. Next, we will scroll down to the `loss_fn` again, shown here:

```
def loss_fn(recon_x, x, mu, logvar):
    BCE = F.binary_cross_entropy(recon_x, x, size_average=False)
    KLD = -0.5 * torch.mean(1 + logvar - mu.pow(2) - logvar.exp())
    return BCE + KLD, BCE, KLD
```

6. The customized loss function in this case is using two methods to measure the divergence between the learned distribution and that observed. The first technique is called *binary cross entropy* (BCE), which measures the difference in the raw input image and the reconstructed one. It's is not unlike how we measured loss in a vanilla AE. The second technique is called Kullback Leibler Divergence (KLD) and is used to measure the actual difference in data distributions.

7. Figure 3-16 shows how KLD can be used to measure the difference between two normal distributions. The shaded area represents the divergence between the two distributions. Our goal of calculating the combined loss using BCE and KLD allows us to learn the values of the mean and standard deviation by minimizing this difference in distributions via the loss function.

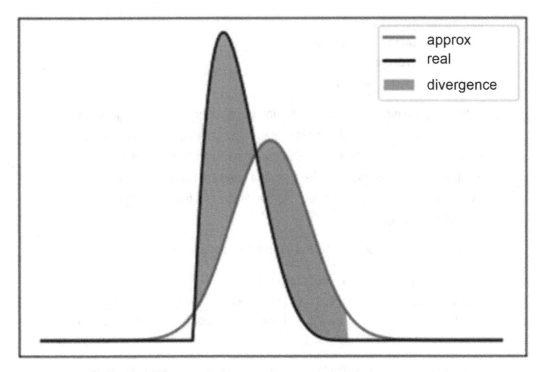

Figure 3-16. *Showing measured divergence between two normal distributions*

8. Let the notebook run until it completes at a minimum of 100 epochs. You will want a good model to better understand the next section.

9. The following block is responsible for generating the example output comparison shown in Figure 3-17. In this figure we see the original images used to input into the VAE to generate the output images. Notice how the output is relatively good considering all the VAE learns is to map the data distribution of the input image.

```
import numpy as np
import matplotlib.pyplot as plt
plt.ion()

import torchvision.utils

def to_img(x):
    x = x.clamp(0, 1)
    return x
```

```python
def show_image(img):
    img = to_img(img)
    npimg = img.numpy()
    plt.imshow(np.transpose(npimg, (1, 2, 0)))

def visualize_output(images, model):
    with torch.no_grad():
        images = images.to(device)
        images, _, _ = model(images)
        images = images.cpu()
        images = to_img(images)
        np_imagegrid = torchvision.utils.make_grid(images[1:50], 10,
        5).numpy()
        plt.imshow(np.transpose(np_imagegrid, (1, 2, 0)))
        plt.show()

images, labels = iter(train_loader).next()

# First visualize the original images
print('Original images')
show_image(torchvision.utils.make_grid(images[1:50],10, 5))
plt.show()

# Reconstruct and visualize the images using the vae
print('VAE reconstruction:')
visualize_output(images, model)
```

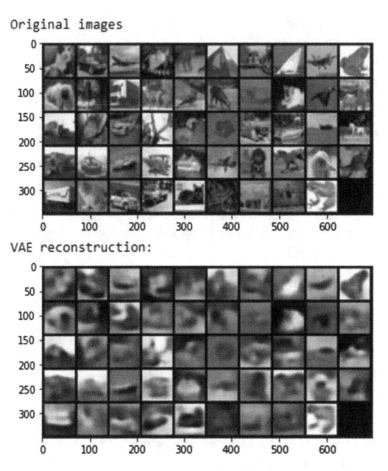

Figure 3-17. *Input images versus output generated images*

With the output in Figure 3-17 you can start to appreciate the power of learning the distribution of data in an image. Realize that with each input image the VAE is learning how the data/pixels in the image are being distributed. It then uses that knowledge captured in the learned mean and standard deviation to create a normal distribution that it then uses to randomly sample from. Thus, each of the output images in the VAE reconstruction is randomly sampled from the learned distribution.

The concept of learning how data is distributed, and mapping that to a known or unknown function, is the core of generative modeling with deep learning. We will use this concept repeatedly as we progress through this book. In the next section, we look at how understanding the distribution of data allows us to also manipulate output.

Variability and Exploring the Latent Space

With a trained VAE model, we can begin to explore the latent space of what the model learns. Throughout this chapter we have been working toward understanding what that latent space looks like. Now we are at a position to unfold that latent space and view the contents visually.

The best way to visualize the latent space encodings in a VAE is to visualize the output or the controlled output. We can do that a couple of ways, and in Exercise 3-6 we explore how to search the output space of a learned VAE.

EXERCISE 3-6. UNDERSTANDING THE VAE

1. Jump to your last and trained version of the GEN_3_conv_VAE_latent. ipynb notebook. If you need to train the notebook again, do so; it can take about up to an hour. We will look at how to save and restore models later, but for now just retrain if you need to.

2. At the bottom of this notebook are two code cells that allow us to map the output of the model either randomly or controlled across the latent space distribution parameters, mean, and standard deviation.

```
with torch.no_grad():
    # sample latent vectors from the normal distribution
    latent = torch.randn(60, 1024, device=device)
    # reconstruct images from the latent vectors
    img_recon = model.decoder(latent)
    img_recon = img_recon.cpu()

    fig, ax = plt.subplots(figsize=(20, 20))
    show_image(torchvision.utils.make_grid(img_recon.data[:100],10,5))
    plt.show()
```

3. The first block of code shown earlier generates the sample latent vectors using the torch.randn function to generate 60 vectors of size 1024. We get 1024 from the latent vector size used in our VAE; 1024 is a much larger latent vector than we used when exploring autoencoders, which is to account for the more complicated training dataset. The rest of the code uses model.decoder to use the latent vector as input to generate a new set of images.

4. Figure 3-18 shows the sample output from this last code cell. The output here effectively represents entirely random vectors fed into the decoder part of the VAE to generate output. As you can see, the results are less than spectacular.

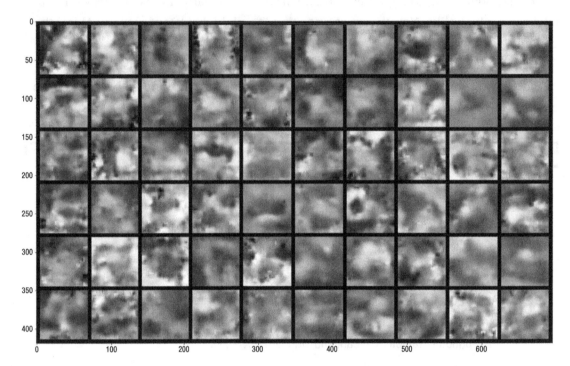

Figure 3-18. *Example output from random latent space generation*

5. The next and last blocks of code demonstrate how we can cycle through the normal distribution parameters to generate not random samples, but samples created from a known distribution. This allows us to understand what the visual area in our model looks like between the data and the latent space.

```
for std in np.arange(.75, 1.5, .1):
  for mean in np.arange(-2, 3, 1):
    with torch.no_grad():
        # sample latent vectors from the normal distribution
        latent = torch.normal(mean, std, size=(10, 1024),
        device=device)
        # reconstruct images from the latent vectors
        img_recon = model.decoder(latent)
        img_recon = img_recon.cpu()
```

```
fig, ax = plt.subplots(figsize=(20, 20))
show_image(torchvision.utils.make_grid(img_recon.
data[:100],10,5))
ax.axis('off')
print(mean, std)
plt.show()
```

6. Now instead of generating an entire batch of random images, we generate a strip of 10 images with each update of the std and mean. This allows us to determine the amount of sensitivity in these parameters but also how they visually generate output.

7. Figure 3-19 shows an extracted snippet of the output from the last code cell. Notice how the change to mean and std change the look of the output. Also notice how the model attempts to replicate features it learned. This feature reconstruction is the result of our convolutional layers creating those features. In some cases, you may see recognizable features that belong to specific classes like cats or dogs. See Figure 3-19.

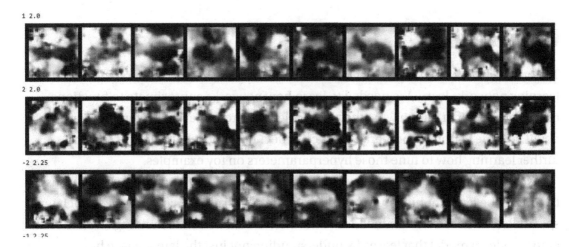

Figure 3-19. *Example output of sample latent space parameters*

8. Now that you understand the how a VAE works, try to tune the hyperparameters batch_size, learning_rate, and epochs. Recall what we learned earlier in this chapter.

9. If you are feeling more advanced, try altering the hidden latent vector size
    h_dim from 1024 to a different number. Also try modifying the z_dim
    hyperparameter from 32 to 64 or 16. What effect does this have? Hint: the code
    where you can find those two hyperparameters is shown here:

```
class ConvVAE(nn.Module):
    def __init__(self, image_channels=3, h_dim=1024, z_dim=32):
```

In the previous exercise, we started to visually explore the latent space in our models.
This technique provides us with greater feedback on how well our model trains and what
it is training to. We will use techniques like we used in the previous exercise repeatedly
throughout this book.

Exploring the latent space in a VAE is useful and provides several insights.
Unfortunately, mapping to distributional parameters still very much limits our model's
learning abilities. Fortunately, the model we already explored, the GAN, already learns
the latent distribution of data. We will explore more about how GANs do this in the next
chapter.

Conclusion

To construct working and practical generative models, we need to understand how deep
learning networks learn. We needed to learn how networks approximate to functions
and what effect that has on what the network learns. From that we explored in some
detail how the various hyperparameters can affect how and what a network learns,
further learning how to tune those hyperparameters on toy examples.

Next, we moved from tuning to understanding how hyperparameters can allow us
to control the latent or hidden space in our models. Learning to control the size of latent
space allowed us to better train the models. Then we advanced to building a variational
autoencoder, a model that learns by understanding not just the latent space but
distributional parameters that map to that latent space.

Finally, from our knowledge of mapping parameters to the latent space, we could
then construct output strictly based on those parameters. This allowed us to see in detail
what our model learns and how sensitive it is to the input data. While we didn't look to
resolve issues with our model in this chapter, we will certainly look to other systems in
the future to improve on it.

In the next chapter, we will revisit GAN and move on to explore its many variations. GANs can be adapted to easily to control the latent space in model generation using several techniques, many of them related to distributional learning. This will allow us to generate content in a more controlled manner.

CHAPTER 4

GANs, GANs, and More GANs

Generative modeling has been around for a few decades, but much of the field didn't start to recognize itself without the discovery of the generative adversarial network (GAN). There is some debate on when GANs were discovered and by whom. One thing is for certain: Ian Goodfellow and his colleagues from the University of Montreal in 2014 deserve a good deal of credit for reinventing the technique of adversarial learning.

GANs are, after all, nothing more than autoencoders that have been split in the middle and flipped around. Goodfellow took this concept a step further by introducing the concepts of the generator (art forger) and the discriminator (art critic) in a true adversarial sense as a way of generating new and novel content. This technique has become so successful in generating new content that there are now hundreds of implementations and variations of this simple model.

In this chapter, we revisit the GAN and start to look at the many implementations that have been adopted as improvements or variations. We start by looking back at convolution in a GAN by improving on the deep convolutional GAN. We will see how to improve on measuring loss or distance by introducing the Wasserstein GAN. Next, we explore what effect discrete data may have on a GAN and how to solve it with the boundary-seeking GAN. We then advance to improving our GAN performance by refining loss measurement with the relativistic GAN. From there, we jump back to understanding and controlling the latent space with the conditional GAN.

GANs have landed at the forefront of generative modeling. They are a technique we will explore tirelessly throughout this book. In this chapter, we start with exploring several base GANs that will provide us with techniques to use later. Here is a quick breakdown of the main topics we will cover in this chapter:

- Feature understanding with the DCGAN

- Unrolling the math of GANs

© Micheal Lanham 2021
M. Lanham, *Generating a New Reality*, https://doi.org/10.1007/978-1-4842-7092-9_4

- Resolving distance with WGAN

- Discretizing boundary-seeking GANs

- Relativity and the relativistic GAN

- Conditioning with CGAN

This is a code-heavy chapter filled with many examples. All these examples use larger training sets, where appropriate, to demonstrate concepts better. Larger datasets could take hours or days to train. While most of the exercises in this chapter may be trained in under an hour, some may take longer. In the next section, we start exploring GANs by understanding the importance of feature learning.

Feature Understanding and the DCGAN

While we already explored the deep convolutional GAN in a previous chapter, what we failed to cover are the details of what really makes this model better. As we already learned, feature extraction of visible features in 2D images can be facilitated with convolutional layers. What we didn't cover in enough detail is how this matters.

In data science we often characterize feature extraction as the process by which we identify known or unknown features in a dataset. While data science uses statistical modeling to extract features, deep learning has numerous ways to do this automatically. Convolution is one of those methods that allows our model to learn what features are important to characterize an object.

Convolution is not the only way to extract features; it is one that works well enough for most image classification and recognition tasks. Regular convolution itself is limited to extracting localized features, that is, features like an eye or nose. Global feature extraction methods, which we will look at in later chapters, could recognize a nose but also relate that the feature needs to be between the eyes.

Since we already covered the DCGAN in Chapter 2, in this exercise we are going to focus specifically on what the model is learning or extracting using convolution. By being able to understand how and what a model learns, we can then derive what it generates. In Exercise 4-1, we revisit the DCGAN with a focus on understanding the features it extracts.

EXERCISE 4-1. DCGAN FEATURE LEARNING

---

1. Open the GEN_4_DCGAN.ipynb notebook from the GitHub project site. If you are unsure how, then consult Appendix B.

2. Run the whole notebook by selecting Runtime ➤ Run all. Then look past the imports cell and on to the first cell with a new class called Hyperparameters:

```
class Hyperparameters(object):
  def __init__(self, **kwargs):
    self.__dict__.update(kwargs)

hp = Hyperparameters(n_epochs=200,
                     batch_size=64,
                     lr=.0002,
                     b1=.5,
                     b2=0.999,
                     n_cpu=8,
                     latent_dim=100,
                     img_size=32,
                     channels=1,
                     sample_interval=400)
```

```
print(hp.lr)
```

3. Hyperparameters is a dictionary helper that allows us to define all the hyperparameters in one place initialized either by a dictionary or by a list of key-value pairs like that shown. Then we can refer to a hyperparameter by using the name hp plus the parameter, as shown in the print(hp.lr) line.

4. The next block of code contains utility code we have reviewed before. After that is a new definition of Generator.

```
class Generator(nn.Module):
  def __init__(self):
    super(Generator, self).__init__()

    self.init_size = hp.img_size // 4
```

```
    self.l1 = nn.Sequential(nn.Linear(hp.latent_dim, 128 * self.
    init_size ** 2))

    self.conv_blocks = nn.Sequential(
      nn.BatchNorm2d(128),
      nn.Upsample(scale_factor=2),
      nn.Conv2d(128, 128, 3, stride=1, padding=1),
      nn.BatchNorm2d(128, 0.8),
      nn.LeakyReLU(0.2, inplace=True),
      nn.Upsample(scale_factor=2),
      nn.Conv2d(128, 64, 3, stride=1, padding=1),
      nn.BatchNorm2d(64, 0.8),
      nn.LeakyReLU(0.2, inplace=True),
      nn.Conv2d(64, hp.channels, 3, stride=1, padding=1),
      nn.Tanh(),
    )

  def forward(self, z):
    out = self.l1(z)
    out = out.view(out.shape[0], 128, self.init_size, self.
    init_size)
    img = self.conv_blocks(out)
    return img
```

5. This class resembles the same class we used for our previous generator. There are a few subtle differences that make this class more abstract and reusable. Note that the vanilla generator in a GAN learns from random noise, while the discriminator is learning from the established, or *ground*, truth.

6. Next, we have the `Discriminator` class that shows a new version of a discriminator for a DCGAN.

```
class Discriminator(nn.Module):
  def __init__(self):
    super(Discriminator, self).__init__()

    def discriminator_block(in_filters, out_filters, bn=True):
      block = [nn.Conv2d(in_filters, out_filters, 3, 2, 1),
               nn.LeakyReLU(0.2, inplace=True),
               nn.Dropout2d(0.25)]
```

```
        if bn:
            block.append(nn.BatchNorm2d(out_filters, 0.8))
        return block

    self.model = nn.Sequential(
        *discriminator_block(hp.channels, 16, bn=False),
        *discriminator_block(16, 32),
        *discriminator_block(32, 64),
        *discriminator_block(64, 128),
    )

    # The height and width of downsampled image
    ds_size = hp.img_size // 2 ** 4
    self.adv_layer = nn.Sequential(nn.Linear(128 * ds_size ** 2, 1),
    nn.Sigmoid())

  def forward(self, img):
    out = self.model(img)
    out = out.view(out.shape[0], -1)
    validity = self.adv_layer(out)

    return validity
```

7. After that we have the cell where the models are defined and the loss function
   is created. Notice how we have the option to use cuda and how the models are
   defined if we are using a GPU.

```
loss_fn = torch.nn.BCELoss()
generator = Generator()
discriminator = Discriminator()

if cuda:
  generator.cuda()
  discriminator.cuda()
  loss_fn.cuda()

# Initialize weights
generator.apply(weights_init_normal)
discriminator.apply(weights_init_normal)
```

8. Move down the rest of the code to the last cell where all the training is defined. Notice how we define the fake and real images with 0.0 or 1.0s a little different than we did before.

```
for epoch in range(hp.n_epochs):
    for i, (imgs, _) in enumerate(dataloader):
        valid = Variable(Tensor(imgs.shape[0], 1).fill_(1.0),
        requires_grad=False)
        fake = Variable(Tensor(imgs.shape[0], 1).fill_(0.0),
        requires_grad=False)
        real_imgs = Variable(imgs.type(Tensor))
        optimizer_G.zero_grad()

        z = Variable(Tensor(np.random.normal(0, 1, (imgs.shape[0],
        hp.latent_dim))))
        gen_imgs = generator(z)
        g_loss = loss_fn(discriminator(gen_imgs), valid)

        g_loss.backward()
        optimizer_G.step()

        optimizer_D.zero_grad()
        real_loss = loss_fn(discriminator(real_imgs), valid)
        fake_loss = loss_fn(discriminator(gen_imgs.detach()),
        fake)
        d_loss = (real_loss + fake_loss) / 2
        d_loss.backward()
        optimizer_D.step()
        batches_done = epoch * len(dataloader) + i
        if batches_done % hp.sample_interval == 0:
            clear_output()
            print(f"Epoch:{epoch}:It{i}:DLoss{d_loss.
            item()}:GLoss{g_loss.item()}")
            visualize_output(gen_imgs.data[:50],10, 10)
```

9. As the notebook runs, watch how the images are generated. Pay particular attention to the blocky nature of how the images are generated. These almost block-like features are a result of the convolutional feature extraction process. If you need to review the image feature extraction process with CNN layers, consult Chapter 2.

Figure 4-1 shows the training process and results of the generated images at various sampling points in training. Notice how the images appear to be stitched together from various patches or what best can be described as *extracted features.*

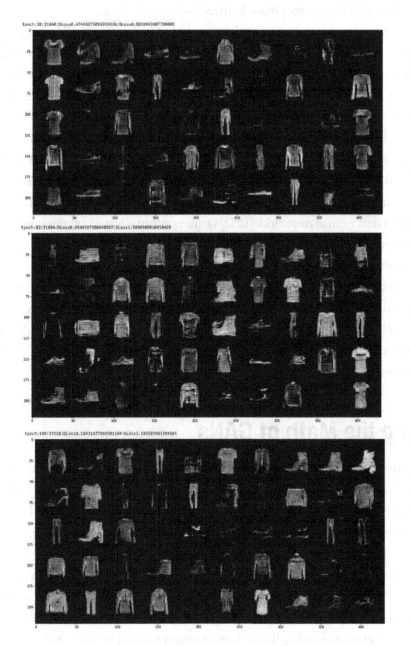

Figure 4-1. *Generated results from the DCGAN at various sampling points*

The dual convolutional GAN was a substantial improvement over the original vanilla GAN due to the ability to perform feature extraction. While visually the results can be impressive, it becomes obvious that the model has limitations. This can be seen with the apparent stitching of the images from learned feature patches.

That doesn't mean convolution or the concept of feature extraction in generative modeling is limited. What we need to consider when using feature extracting layers like convolution or recurrent layers are the details and context of the extraction. In convolution, the detail is limited to the minimum patch size of our smallest convolutional layer, and the context is always localized.

Recurrent layers are a special form of deep learning layers that can learn or extract sequences in data. These types of layers are often used to extract features in time-series data or natural language text analysis. Conceivably, they could be used to extract features from video data or sequences of data, but this has not been shown to be an efficient solution. Recurrent neural networks (RNNs) are computationally expensive, and other solutions are now used to perform the same type of feature extraction. We won't cover RNNs any further in this book.

Since convolution can be somewhat limited as applied to generative modeling, we will look at various other options that can provide much better results. Often, though, the results themselves will often be dictated by the input data you are trying to simulate. In the next section, we jump back into the math of GANs to understand what those better options may be.

Unrolling the Math of GANs

To better understand how GANs learn, it is important to understand the math or at the least the intuition behind the math. Fortunately, we already looked at some of those math basics in Chapter 3 when we discussed the variational autoencoder (VAE). If you recall, a VAE learns by understanding and modeling the input data distribution. Then it generates samples from that learned distribution.

As it turns out, the mathematics of a GAN are like a VAE in that they learn by understanding the distribution they are trying to discern or generate. However, the math we use to get there is derived in a slightly different manner. It is important to understand how GANs work at a low level to fix or resolve issues when later training.

We will start at the basic loss function we use in a GAN, the binary cross-entropy loss function, as shown here:

$$L(\check{y}, y) = [y \times log(\check{y}) + (1 - y) \times log(1 - \check{y})]$$

where:

$y = originaldata$

$\check{y} = generateddata$

To optimize this function, let's first look at determining the loss for the discriminator. When training the discriminator, we assume for real data, $y = 1$. Conversely, for the generated data, we let $\check{y} = D(x)$; then substituting in the last equation, we get loss for the discriminator with the following:

$$L\big(D(x), 1\big) = log\big(D(x)\big)$$

Then for output generated from the generator, we assume $y = 0$ (fake data) and then $\check{y} = D(G(z))$, where z represents random sample vector space. Substituting back into the loss equation, we get loss for the generator with the following:

$$L\big(D(G(z)), 0\big) = log\big(1 - D(G(z))\big)$$

Then we can combine the loss for both equations and maximize to determine the total discriminator loss with this:

$$L^{(D)} = max\Big[log\big(D(x)\big) + log\big(1 - D(G(z))\big)\Big]$$

Since the generator is competing against the discriminator, its job is to perform the opposite, and thus, we calculate the minimal loss for the generator with the following:

$$L^{(G)} = min\Big[log\big(D(x)\big) + log\big(1 - D(G(z))\big)\Big]$$

To simplify this view, we can combine both of those equations using the shorthand equation shown here:

$$L = \underset{G \quad D}{min\,max}\left[log\left(D(x)\right) + log\left(1 - D\left(G(z)\right)\right)\right]$$

Now the previous loss function only defines the amount of loss over a single pixel or data point. To cover a whole image or set of data, we need to unfold the equation to match the original GAN equation from the paper on GANs by Ian Goodfellow and colleagues:

$$\underset{G \quad D}{min\,max}\,V(D,G) = \underset{G \quad D}{min\,max}\left(E_{xPdata(x)}\left[log\left(D(x)\right)\right] + E_{zP(z)}\left[log\left(1 - D\left(G(z)\right)\right)\right]\right)$$

where:

$E_{xPdata(x)}$ is the expectation or distribution of real data.

$E_{zP(z)}$ is the expected distribution of fake data.

This means we are trying to optimize the expected generated distribution to match the real or actual data distribution. Again, this is not unlike our previous look at VAEs where the goal was to optimize the sampling distribution. We will often see the equation rewritten as follows:

$$E_x\left[log\left(D(x)\right)\right] + E_z\left[1 - log\left(D\left(G(z)\right)\right)\right]$$

Now the problem becomes how well our generator can learn to model the real data distribution. However, in practice, if the divergence in between the generated distribution and the real distribution becomes too large, the generator will stall and suffer from a vanishing gradient problem.

Figure 4-2 shows how the expected generated distribution and the discriminators' expectation of the real distribution may converge or diverge during training. As the discriminator gets better at identifying fake and real images, this may increase divergence between the generated expected distribution. Likewise, if the generator is poorly configured, it may start out with a poor expectation or not be able to learn the expected distribution.

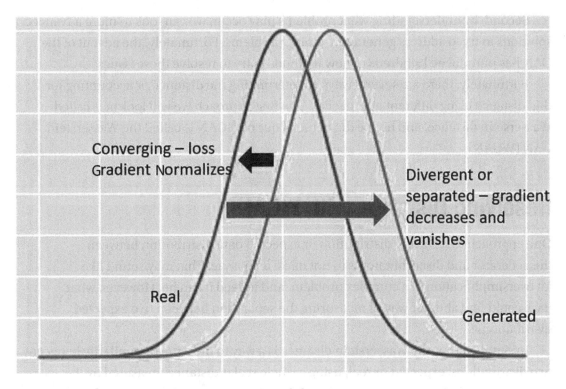

Figure 4-2. Understanding generator and discriminator expectation

If the expected real and generated distributions become too diverse or separate with no overlap, then the generator will suffer from vanishing gradient loss. Vanishing gradients are an issue in that the loss gradient becomes so small that it has no effect on training the model. What we see is the generator model will just stall, not making any progress.

The mathematics behind why the loss vanishes are beyond the scope of the book. However, the intuition here is that as the expected real and generated distributions diverge with no overlap, this becomes problematic. We have two important concepts to take away from this knowledge.

First, our generator and discriminator models need to be trained in tandem with neither gaining clear superiority over the other. The better the expected and generated distribution matches, the better our models will train. A discriminator too good at identifying fakes versus real early on makes the challenge of the generator insurmountable.

Second, by understanding what problems may occur, we can look to more advanced solutions to try to address generator training problems. Fortunately, the advent of the GAN has introduced hundreds of new methods to try to resolve these issues.

Fortunately, there are several ways of determining the distance or accounting for that distance using different approaches. The first approach we will look at is called Wasserstein distance, and hence the name of our next GAN is called the Wasserstein GAN (WGAN).

Resolving Distance with WGAN

One approach to resolving distribution or expected data distribution between the generator and discriminator is to not make it an issue. That may sound like an oversimplification of a complex problem, and indeed it can be. However, what if we could find another way of measuring the separation between two expected distributions?

In simple terms, the Wasserstein distance is a way to measure dissimilarity between two distributions, expected or real, using a single scalar distance measured by the amount of work to transform one into another. The approach often used to describe this is to think of two piles of dirt. Both piles are of equal mass/size but are different shapes. The Wasserstein metric or distance is the amount of work it would take to transform the first pile of dirt into the second, as shown in Figure 4-3.

That means we ignore the distance between the two distributions, and this allows us to greatly simplify the math. Since we no longer worry about the distance, the loss functions become simplified and in turn are the difference between real and fake. It also means that the discriminator can no longer discriminate between what is real or fake, but instead just critique. As such, we now call our discriminator a *critic*.

This method of measuring the distance between two distributions is also known as *earth mover's distance*. This term denotes the amount of work it would take to move material/dirt from one pile to another using discrete amounts. In simple terms, you could think of how many trips a truck would need to move a pile of dirt from one location to the next.

The math for all this breaks down the loss functions for the critic and generator to just be the following:

Critic Loss: $D(x) - D(G(z))$

Generator Loss: $D(G(z))$

The other key difference here is the generator tries to maximize the function, whereas in a vanilla GAN the generator minimizes loss.

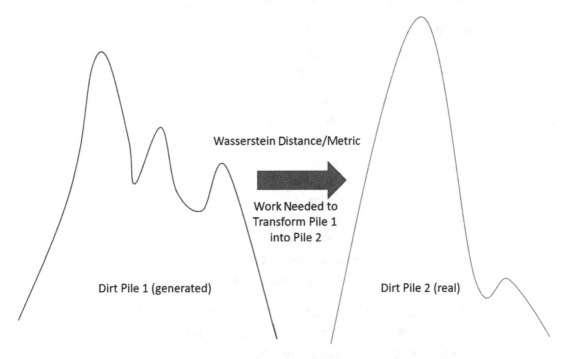

Figure 4-3. *Wasserstein distance explained*

There are a couple of other key differences that result from these assumptions. First, the weights in the network need to be clamped into a certain range to avoid vanishing/ exploding gradients in the critic/discriminator. Second, we need to increase training iterations of the critic so that it works to approximate the real distribution sooner.

Now that we have some understanding of the concepts of how this works, let's dive into another code example and see how it works in practice. In Exercise 4-2, we look at an implementation of the WGAN trained from the MNIST fashion dataset. It is important that we learn the baseline training datasets to better understand differences in GANs.

EXERCISE 4-2. EXPLORING THE WGAN

1. Open the GEN_4_WGAN.ipynb notebook from the GitHub project site. If you are unsure how, then consult Appendix B.

2. Run the whole notebook by selecting Runtime ➤ Run all. Then look past the imports cell and on to the first cell with the Hyperparameters class, and examine the new hyperparameters, n_critic and clip_value:

```
class Hyperparameters(object):
  def __init__(self, **kwargs):
    self.__dict__.update(kwargs)

hp = Hyperparameters(n_epochs=200,
                     batch_size=64,
                     lr=0.00005,
                     n_cpu=8,
                     latent_dim=100,
                     img_size=32,
                     channels=1,
                     n_critic=25,
                     clip_value=.005,
                     sample_interval=400)
```

3. The new hyperparameters are n_critic and clip_value. n_critic defines the number of critic iterations in training, and clip_value sets the limits of the weights in the critic/discriminator.

4. You can scroll down to the last code block, which is the training block. Most of the code in this example we covered in previous examples, so we don't need to review it here.

5. We will focus on the inner block of training code starting with the discriminator/critic loss, as shown here:

```
valid = Variable(Tensor(imgs.shape[0], 1).fill_(1.0), requires_
grad=False)
fake = Variable(Tensor(imgs.shape[0], 1).fill_(0.0), requires_
grad=False)

real_imgs = Variable(imgs.type(Tensor))

optimizer_G.zero_grad()

z = Variable(Tensor(np.random.normal(0, 1,
      (imgs.shape[0], hp.latent_dim))))

fake_imgs = generator(z).detach()
```

```
d_loss = -torch.mean(discriminator(real_imgs))
        + torch.mean(discriminator(fake_imgs))

d_loss.backward()
optimizer_D.step()
```

6. The top part of that generates the ground truths of valid and fake tensors we have seen before. After that, we see the set of `real_imgs` extracted from the batch as well as `fake_imgs` generated from a random z. After that, the critic/discriminator loss is calculated from subtracting the mean of passing `real_imgs` into the discriminator from the mean of the `fake_imgs` passed into the discriminator. In a vanilla GAN, we would use the binary cross-entropy function to measure the loss.

7. Next, we will look at the code that does the weight clamping. We again need to clamp the weights into a narrow window to avoid exploding gradients.

```
for p in discriminator.parameters():
        p.data.clamp_(-hp.clip_value, hp.clip_value)
```

8. After that, we start the section that controls the iteration of the generator loss starting with the `if` statement:

```
if i % hp.n_critic == 0:
        optimizer_G.zero_grad()
        gen_imgs = generator(z)

        g_loss = -torch.mean(discriminator(gen_imgs))

        g_loss.backward()
        optimizer_G.step()
```

9. The `if` statement controls how after the generator is run per critic training pass. After that, most of the code is familiar, but notice the simplification for calculation of loss, `g_loss`, for the generator.

The output from this training example will be somewhat disappointing considering our most recent expeditions into running more advanced GANs like the DCGAN. It becomes obvious that the WGAN is not well suited or equipped to learn the MNIST fashion dataset. If we used a different dataset, we would expect much better results. So, what is wrong with the MNIST fashion dataset?

The problem with the fashion dataset is the makeup of the images are too diverse to easily learn a common or generalized expected distribution. In other words, the picture of a shoe is far too different than a picture of a sweater or pants. This problem becomes compounded with the more classes or domains of data we try to train on.

While we can fix the problem of learning diverse domains by doing feature extraction, like we did with the DCGAN, ideally we want to solve the root cause of the problem by looking how we can manage or characterize loss across domains or classes. We will look at one such approach in the next section.

Discretizing Boundary-Seeking GANs

We can characterize visual differences in images or other data as classes. In the MNIST fashion dataset there are 10 classes that have similarities and key differences. Figure 4-4 shows the differences between the classes of data in the fashion set.

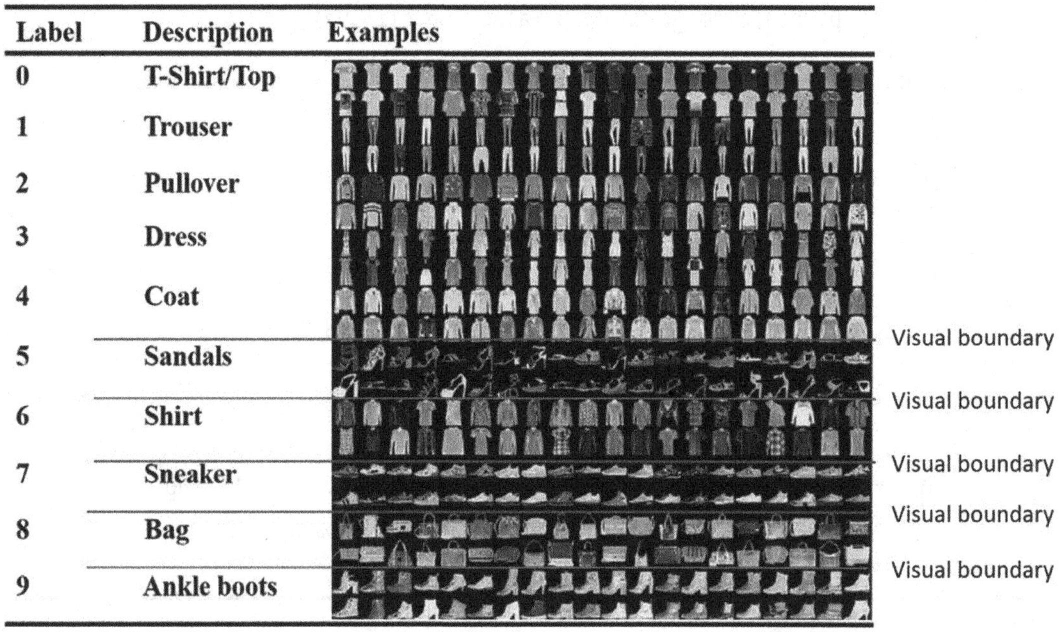

Label	Description	Examples	
0	T-Shirt/Top		
1	Trouser		
2	Pullover		
3	Dress		
4	Coat		
5	Sandals		Visual boundary
6	Shirt		Visual boundary
7	Sneaker		Visual boundary
8	Bag		Visual boundary
9	Ankle boots		Visual boundary

Figure 4-4. *Visualizing the boundaries in the fashion MNIST data*

In Figure 4-4, you can clearly see how some classes are visually similar, while other classes are not. If we reduced our training dataset to just the first five classes (T-shirt, Trouser, Pullover, Dress, and Coat), our model generation would be simpler.

For this example, we are equating our visual perception of an image to an expected distribution. That generally works well, provided the data can be easily generalized. If we look at the Sandals group from Figure 4-4, notice the visual detail for each image. Detail can also equate to more complicated expected distributions to learn.

The other reason this causes an issue for our GAN is because of the way deep learning networks learn through calculus. By using calculus, our input data should always be continuous. That is, it must show common data that can be easily generalized and that can be transformed across images. In the case of the fashion MNIST dataset, we can clearly see many places where transforming one image like sandals into a pullover would not be a continuous transition.

We can look to solve these shortcomings in GANs by again looking at how we calculate loss. One such idea was proposed in a paper titled "Boundary-seeking Generative Adversarial Networks" by R Devon Hjelm and colleagues. The idea they proposed was to measure expected distribution differences using importance weights.

Importance weights or weighing is a method by which the more important features of an image or dataset are given more strength or output. This has the effect of isolating those features important to a given class or classes, thus allowing the model to learn more discrete or varied sets of class data.

Using importance sampling allows the derivation of a policy gradient solution to better approximate the loss. Importance sampling is a technique for estimating the parametric parameters required to regenerate a distribution. Policy gradient methods have their background in reinforcement learning and are simply a way of optimizing loss using parameterized solutions.

Reinforcement learning is a method that teaches models also called *agents* to learn using rewards. Like unsupervised learning, the agent can learn on its own through trial-and-error exploration. It is possible this form of learning could be used for future generative modeling solutions but is not something we will explore in this book. Policy gradient methods are a subset of RL methods that attempt to converge a policy using gradient clipping.

The differences between the BGAN and a vanilla GAN are subtle and best explored by running some code and looking at the results. In Exercise 4-3, we will do just that and look to use a BGAN again on the MNIST fashion dataset.

EXERCISE 4-3. BREAKING BOUNDARIES WITH BGAN

1. Open the `GEN_4_Boundary_Seeking_GAN.ipynb` notebook from the GitHub project site. If you are unsure how, then consult Appendix B.

2. Run the whole notebook by selecting Runtime ➤ Run all. Then look past the `imports` cell and on to the first cell with the `Hyperparameters` class, and examine the hyperparameters, all of which we have seen before.

3. The only major change is the determination of loss and how the functions are defined, as shown here:

```
def boundary_seeking_loss(y_pred, y_true):
    """

    Boundary seeking loss.
    Reference: https://wiseodd.github.io/techblog/2017/03/07/boundary-
    seeking-gan/
    """
    return 0.5 * torch.mean((torch.log(y_pred)
      - torch.log(1 - y_pred)) ** 2)

d_loss_fn = torch.nn.BCELoss()

generator = Generator()
discriminator = Discriminator()

if cuda:
  generator.cuda()
  discriminator.cuda()
  d_loss_fn.cuda()
```

4. We can see the definition of the `boundary_seeking_loss` function and the equation used to determine that loss. That equation is shown here:

$$\frac{1}{2} \times mean$$

5. This equation becomes the loss for the generator. Notice how `y_true` or x is not used in the equation. Again, this has to do with the policy gradient method, which looks at predictions and not actuals.

6. Scrolling down to the training block of code, we can see how this new loss equation is used in the generator training.

```
# Generator loss
gen_imgs = generator(z)

# measure ability to fool discriminator
g_loss = boundary_seeking_loss(discriminator(gen_imgs), valid)

g_loss.backward()
optimizer_G.step()

#  Discriminator loss
optimizer_D.zero_grad()

real_loss = d_loss_fn(discriminator(real_imgs), valid)
fake_loss = d_loss_fn(discriminator(gen_imgs.detach()), fake)
d_loss = (real_loss + fake_loss) / 2
```

7. The highlighted line shows the only code change in the training loop. With that one line of code and subtle change to our loss calculation, we can alter the results substantially.

Figure 4-5 shows the results of training after only a couple of epochs, and it can clearly be seen that the BGAN is able to identify classes of clothing and start to generate them. As opposed to the WGAN, which struggled initially to find common ground, the BGAN can almost immediately distinguish between classes.

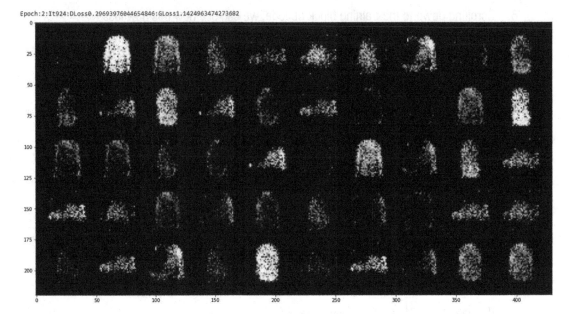

Figure 4-5. *Early results from running the BGAN on fashion MNIST*

However, after fully training the BGAN, the results are still not entirely as good as we had hoped. This has more to do with the diverse classes and the amount of detail specific to some classes like sandals. In fact, if you consider the final training output, you will note that those specialized classes (like sandals) are poorly represented.

We should expect with more longer-term training to eventually get good results, but as you can see, there are still limitations on this model. The boundary-seeking GAN was developed to work across data with discrete boundaries and could also be well suited to other discrete forms of data such as tabular datasets that show discontinuous data.

As we have seen so far, a lot can go into the loss functions for both the generator and the discriminator. In the next section, we look at another approach to improving the loss function in a GAN.

Relativity and the Relativistic GAN

We have looked at two distinct methods approaches to determining loss in a GAN. With a vanilla or standard GAN, loss is measured in absolute terms. In a WGAN, we learned how loss was determined in a more relativistic approach using an earth mover's algorithm. By using a more relative approach to calculating loss, the WGAN was able to account for some training deficiencies we had seen earlier in the standard GAN.

The relativistic GAN is another approach that uses a relative method to calculate the distance or loss in a GAN. In a standard GAN, the discriminator estimates the probability that real data is real, and the generator's job to increase the probability that fake data is also seen as real. However, the authors of the RGAN also want to account for the generator, reducing the probability of real images being real.

Accounting for prior knowledge that half the data it sees is always fake and the probability fake data becomes more real allows the GAN to infer if an image is real or fake using previously observed observations. The result is a relativistic discriminator that can account for how the expected distribution is changing over time through learning.

This alters the loss function of the generator to resemble the following:

$$L^{(G)} = logit\big(D\big(G(z)\big) - \big(D(x)\big), 1\big)$$

where:

$logit$ = the binary cross entropy logit loss function $log\left(\dfrac{x}{1-x}\right)$.

The *logit* function is often described as the odds function, and it returns the odds of the given probability of expectation the result is real or fake.

The relativistic discriminator is fed two combined versions of loss, the real and fake given by the following:

$$L^{(D)}_{real} = logit\big(\big(D(x)\big) - D\big(G(z)\big), 1\big)$$

$$L^{(D)}_{fake} = logit\big(D\big(G(z) - \big(D(x)\big)\big), 0\big)$$

$$L^{(D)}_{total} = \dfrac{\big(L^{(D)}_{real} + L^{(D)}_{fake}\big)}{2}$$

Using the logit function turns the output from an expected probability to the expected odds. In other words, do not guess what the real or fake data is but rather the probability or odds that data is real or fake. The idea is to think of how far the answer is likely from the truth given past views of the data.

The authors of the RGAN also proposed another method that compares calculated odds against average/mean odds from the real or predicted data. They called this variation the *relativistic average GAN* (RaGAN).

Now that we understand how the loss works a little better, we can look at a concrete implementation in PyTorch. In Exercise 4-4, we look at employing an RGAN and RaGAN trained on the fashion MNIST dataset.

EXERCISE 4-4. RELATING THE RELATIVISTIC GAN

1. Open the GEN_4_Relativistic_GAN.ipynb notebook from the GitHub project site. If you are unsure how, then consult Appendix B.

2. Run the whole notebook by selecting Runtime ➤ Run all. Then look past the imports cell and on to the first cell with the Hyperparameters class and examine the hyperparameters, all of which we have seen before. There is a new hyperparameter called rel_avg_gan, which controls if the GAN is run as an RGAN or RaGAN.

3. This GAN uses convolution for additional feature extraction. The generator is similar to previous looks at the DCGAN, but the discriminator is constructed differently, as shown here:

```python
class Discriminator(nn.Module):
    def __init__(self):
        super(Discriminator, self).__init__()

        def discriminator_block(in_filters, out_filters, bn=True):
            block = [nn.Conv2d(in_filters, out_filters, 3, 2, 1),
                nn.LeakyReLU(0.2, inplace=True), nn.Dropout2d(0.25)]
            if bn:
                block.append(nn.BatchNorm2d(out_filters, 0.8))
            return block

        self.model = nn.Sequential(
            *discriminator_block(hp.channels, 16, bn=False),
            *discriminator_block(16, 32),
            *discriminator_block(32, 64),
            *discriminator_block(64, 128),
        )
```

```
        # The height and width of downsampled image
        ds_size = hp.img_size // 2 ** 4
        self.adv_layer = nn.Sequential(nn.Linear(128 * ds_size ** 2, 1))

    def forward(self, img):
        out = self.model(img)
        out = out.view(out.shape[0], -1)
        validity = self.adv_layer(out)

        return validity
```

4. Notice how we define an inner function called `discriminator_block`, which sets up the convolutional layers.

5. Now jump to the training loop and look at how the inner code calculates the loss with our updated equations.

```
# Generator
optimizer_G.zero_grad()

z = Variable(Tensor(np.random.normal(0, 1, (imgs.shape[0], hp.latent_
dim))))

gen_imgs = generator(z)

real_pred = discriminator(real_imgs).detach()
fake_pred = discriminator(gen_imgs)

if hp.rel_avg_gan:
    g_loss = loss_fn(fake_pred - real_pred.mean(0, keepdim=True),
   valid)
else:
    g_loss = loss_fn(fake_pred - real_pred, valid)

g_loss.backward()
optimizer_G.step()

# Discriminator
optimizer_D.zero_grad()
```

```
real_pred = discriminator(real_imgs)
fake_pred = discriminator(gen_imgs.detach())

if hp.rel_avg_gan:
    real_loss = loss_fn(real_pred - fake_pred.mean(0, keepdim=True),
    valid)
    fake_loss = loss_fn(fake_pred - real_pred.mean(0, keepdim=True),
    fake)
else:
    real_loss = loss_fn(real_pred - fake_pred, valid)
    fake_loss = loss_fn(fake_pred - real_pred, fake)

d_loss = (real_loss + fake_loss) / 2

d_loss.backward()
optimizer_D.step()
```

6. Let the example train to completion and then go back and switch the `rel_avg_gan` hyperparameter to false or true. Then compare the results of the RGAN and the RaGAN.

Figure 4-6 shows the early results of training the RGAN on the fashion dataset. As you can see, the results are reasonably impressive compared to our previous attempts. Notice how the feature patching we observed with the DCGAN has also been cured. You can also pick up fine details on clothing items like those problem sandals.

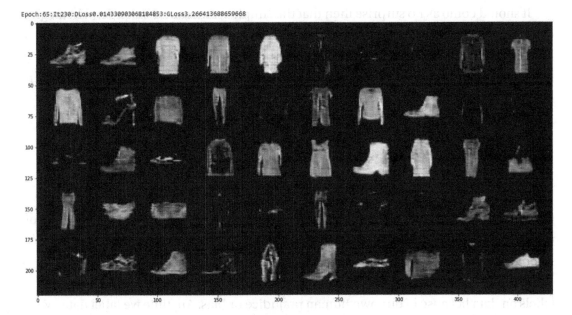

Epoch:65:It230:DLoss0.014330903068184853:GLoss3.266413688659668

Figure 4-6. *Early training results on the RGAN*

So far, we looked at a few approaches in GANs using various loss functions in the hopes of improving the results. We looked at changing the perspective of the loss function from absolute to relative in the WGAN and RGAN/RaGAN, and we looked at how to better classify discrete or out-of-bounds data with a boundary-seeking GAN. In the next section, we look at another variation of the GAN that improves on domain recognition and attribution.

Conditioning with CGAN

As we have seen, there are several variations to account for loss or expected loss in a GAN. All these methods used an expectation that the GAN would learn the entirety of the image or other data without assistance. This requires the GAN to not only learn what is fake or real but also learn latent domains of data.

A latent domain is a hidden domain with respect to the GAN or other model. The GAN needs to learn this domain on its own to generate realistic images for that domain or class. As an example, consider the MNIST fashion dataset we have been GAN training all chapter. This dataset consists of 10 classes, some more unique than others, that the GAN also must learn. We see what issues this caused when we compared the results from looking at the more diverse and detailed classes, like sandals.

It should come as no surprise then that the simple workaround or helper we can give to a GAN is to feed the labels in with the data. So, we tell the GAN not only that it has this image to learn but also what class or domain it belongs to. By doing that, we no longer need to change our loss function, just the inputs into the general loss function, as shown here:

$$L = \frac{\min}{G} \frac{\max}{D} \left[\log\big(D(x, label)\big) + \log\big(1 - D\big(G(z, label)\big)\big) \right]$$

The big change here is that we now also feed the label into the discriminator and generator along with the data. By feeding the label into the models, we now take some of the difficulty away from the model to learn a domain or class on its own. Instead, we give it some help by telling it the class we have tagged it with.

For pure AI or generative modeling, ideally we do not want to have to provide labels on data because of our own human prejudice or bias. Anytime we label data, we are putting a human bias on that data. This is the reason we often prefer to feed deep learning models with raw, unlabeled data and let the model learn on its own. In a vanilla GAN, we essentially always have a minimum of real/fake labeling of data.

Adding labels into our GAN upgrades it to a conditional GAN (CGAN). We call it *conditional* since we provide the conditions or labels of the domains to the CGAN. For our purposes, we are going to look at a cDCGAN or conditional DCGAN in Exercise 4-5.

EXERCISE 4-5. RELATING THE RELATIVISTIC GAN

1. Open the GEN_4_cDCGAN.ipynb notebook from the GitHub project site. If you are unsure how, then consult Appendix B.

2. Run the whole notebook by selecting Runtime ➤ Run all. Then look past the imports cell and on to the first cell with the Hyperparameters class and examine the new value n_classes = 10. This sets the number of classes to be fed as labels into the model.

3. Scroll down to the Generator class definition, as shown here:

```
class Generator(nn.Module):
  def __init__(self):
    super(Generator, self).__init__()
```

```
    self.label_emb = nn.Embedding(hp.n_classes, hp.n_classes)

  def block(in_feat, out_feat, normalize=True):
    layers = [nn.Linear(in_feat, out_feat)]
    if normalize:
        layers.append(nn.BatchNorm1d(out_feat, 0.8))
    layers.append(nn.LeakyReLU(0.2, inplace=True))
    return layers

  self.model = nn.Sequential(
    *block(hp.latent_dim + hp.n_classes, 128, normalize=False),
    *block(128, 256),
    *block(256, 512),
    *block(512, 1024),
    nn.Linear(1024, int(np.prod(img_shape))),
    nn.Tanh()
  )

def forward(self, noise, labels):
  gen_input = torch.cat((self.label_emb(labels), noise), -1)
  img = self.model(gen_input)
  img = img.view(img.size(0), *img_shape)
  return img
```

4. Notice in the `forward` function how the label is concatenated with the random
    noise. The label embeddings are learned using a special layer called an
    embedding layer. Embedding layers are like autoencoders except the output is
    the middle-learned embedding.

5. Next, we will jump down to the `Discriminator` class definition and see how
    the labels are fed in the same way using the learned embeddings.

```
class Discriminator(nn.Module):
  def __init__(self):
    super(Discriminator, self).__init__()

    self.label_embedding = nn.Embedding(hp.n_classes, hp.n_classes)

    self.model = nn.Sequential(
        nn.Linear(hp.n_classes + int(np.prod(img_shape)), 512),
        nn.LeakyReLU(0.2, inplace=True),
```

```
        nn.Linear(512, 512),
        nn.Dropout(0.4),
        nn.LeakyReLU(0.2, inplace=True),
        nn.Linear(512, 512),
        nn.Dropout(0.4),
        nn.LeakyReLU(0.2, inplace=True),
        nn.Linear(512, 1),
    )

    def forward(self, img, labels):
        d_in = torch.cat((img.view(img.size(0), -1), self.label_
        embedding(labels)), -1)
        validity = self.model(d_in)
        return validity
```

6. From here, we are familiar with the rest of the code, so we can skip down to the training code and look at the specific sections.

```
# Generator
z = Variable(FloatTensor(np.random.normal(0, 1, (batch_size,
hp.latent_dim))))

gen_labels = Variable(LongTensor(np.random.randint(0, hp.n_classes,
batch_size)))

gen_imgs = generator(z, gen_labels)

validity = discriminator(gen_imgs, gen_labels)
g_loss = loss_fn(validity, valid)

# Discriminator
validity_real = discriminator(real_imgs, labels)
d_real_loss = loss_fn(validity_real, valid)

validity_fake = discriminator(gen_imgs.detach(), gen_labels)
d_fake_loss = loss_fn(validity_fake, fake)
```

7. The only major change here is now we push the labels as inputs into both the discriminator and the generator.

8. As you let this sample run and watch the results, are they what you expected? Could we add other relative forms of loss like the WGAN or RGAN/RaGAN?

When this sample completes, it will unfortunately be obvious that we still have more work to do. However, notice how this model can create detailed images from those difficult classes like sandals. In some cases, now those difficult classes start to look more realistic than our previous attempts.

At this point, you could go back and look at combining the loss from a WGAN or RGAN/RaGAN into a CGAN or cDCGAN. Some of this has already been done in plenty of other GAN variations you can search for. For us, though, we are going to move on and look at more advanced techniques around generative modeling in later chapters. Now, though, let's finish up the chapter with our conclusions in the next section.

Conclusion

As we saw in this chapter, there are several GAN variations that try to increase performance by looking at loss from different perspectives. From the standard/vanilla GAN, we saw how performance can be improved with convolution and feature extraction. We also look at the Wasserstein and relativistic GANs that accounted for loss in relative terms. Then we tried to account for learning domains or classes in data with the boundary and conditional GANs.

Throughout this chapter we also studied the loss equations and how they varied from one variation to another. By digging under the covers and studying those equations, we began to understand how loss can be learned in absolute or relative terms, as well as looked at ways to improve on those harder-to-learn domains or classes.

Ultimately, what we hope to take away from this chapter is the knowledge that GANs are susceptible to loss and how loss is calculated. Understanding this gives us the power to pick the right loss or set of loss equations that may be relevant to our datasets. After all, if you are building a GAN specialized on your data, you will always want to try several GAN variations and pick the best for your needs.

In the next chapter, we will move on from understanding loss in GANs to exploring variations in training.

CHAPTER 5

Image to Image Content Generation

In days of old, before cameras captured everything, if a crime were committed and it was particularly heinous, the police would send a sketch artist around to interview witnesses. The purpose of this was to build an image or likeness of the criminal and expedite an arrest.

Sketch artists would work by questioning witnesses about features, and the witness would answer yes or no to help the artist create a better likeness. This process is not unlike that of a GAN in that the artist is the generator and the witness is the discriminator. Over time, instead of relying on verbal descriptions, the sketch artists carried along visual aids to help the witness match features.

Sticking with the sketch artist analogy, we have come to the point that simple text/ label descriptions will go only so far identifying the features we want generated. Now we want to provide the ability to demonstrate important features or characteristics the generators should focus on through using images.

For this chapter, we move on from using attributes to control important feature generation to using actual images. Before jumping into the GANs, we will look at how we can perform image to image translation using an image segmentation built around a U network (UNet). We will uncover how a UNet can extract and learn features.

Then we will move on to a couple of examples of GANs that function by translating one image into another. We start with Pix2Pix for understanding the translation of images with a GAN. Then we move on to DualGAN and explore how a dual architecture can enhance learning.

Using image to image pairings is successful in training image translation models, but they often fall short of understanding translation variety. That is, one image could be possibly translated into multiple correct variations. Not unlike translating one phrase

© Micheal Lanham 2021
M. Lanham, *Generating a New Reality*, https://doi.org/10.1007/978-1-4842-7092-9_5

to another in languages, there are often multiple possible translations. We will use BicycleGAN to understand how multiple translations of an image may be possible.

Finally, we finish the chapter by breaking the paired image requirement, and we look at our first GAN, DiscoGAN, for performing image domain translation. We look at several new interesting datasets and train models to translate apples to oranges, horses to zebras, or works of art by Monet to photos.

To summarize, this is what we will be looking at in this chapter:

- Segmenting images with a UNet

- Translating images with Pix2Pix

- Seeing double with DualGAN

- Riding the latent space on BicycleGAN

- Discovering domains with DiscoGAN

Being able to generate content that contains relevant features allows us to control generators to generate a reality we define. In this chapter, we take our next steps into becoming generative gurus and look to augment generator training with additional image characteristics. In the next section, we look at a key element of extracting features from images: a UNet.

Segmenting Images with a UNet

Convolutional network layers are excellent at extracting localized features in images, but their ability to reconstruct those same features is less than ideal. If you recall, we have seen this a few times now when exploring generative models that used CNNs. These models often poorly rebuilt the features, and the image typically looks like nothing more than a patchwork. To resolve issues such as these in generators, we need to look beyond simple CNNs.

One such extension to CNNs and feature rebuilding was developed with a UNet, as shown in Figure 5-1. The architecture derives its name from the shape of the model as it looks like a *U*. Inside, convolutional layers are still used, but instead of trying to rebuild images from a convolutional transpose, the image is rebuilt from the actual learned feature mapping.

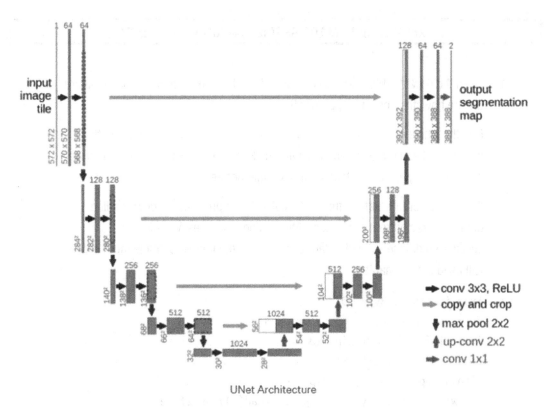

Figure 5-1. *UNet architecture*

If you recall when working with CNNs in our models, we will often use a
`ConvTranspose2D` layer when rebuilding an image. The problem with this is that learned
transposed features are different than the original features. The UNet architecture allows
the model to use the same features as used when classifying or discriminating the image
to regenerate or segment it.

Image segmentation is a key prerequisite to understanding how to extract and
rebuild key features. Remember those sketch artists? Well, they advanced to segmenting
out parts of the face and using those parts to rebuild new faces. This is indeed very much
what we will be doing when we start to translate or pair images to images for generation.

Before we get to that, though, we should probably back up and understand in
practice how the UNet works and how it can be used to extract or segment images. So in
Exercise 5-1, we will build a UNet model to extract and segment images of a fish. As we
go through this exercise, we will talk about the modifications of the architecture and how
loss is calculated.

137

EXERCISE 5-1. IMAGE SEGMENTATION WITH A UNET

---

1. Open the GEN_5_UNet.ipynb notebook from the GitHub project site. If you are unsure how, then consult Appendix B.

2. Run the whole notebook by selecting Runtime ➤ Run all. Then look past the imports cell and on to the first cell with the Hyperparameters class. We have seen all the hyperparameters in this example before.

3. Past the hyperparameters, the next block of code provides for downloading datasets directly from Dropbox or other online resources. We will use specialized datasets now that are designed for the image to image models we will build in this chapter.

```
from io import BytesIO
from urllib.request import urlopen
from zipfile import ZipFile
zipurl = hp.dataset_url
with urlopen(zipurl) as zipresp:
    with ZipFile(BytesIO(zipresp.read())) as zfile:
        zfile.extractall(image_folder)
        print(f"Downloaded & Extracted {zipurl}")
```

4. After this block of code is a new class called FishDataset. As we move to more advanced data, we will often need to extend DataSet to accommodate loading those new sets. Extending Dataset allows us to fine-tune the way Dataloaders load data. We will extend the Dataset class in this manner throughout this and future chapters. The following code shows the start of the class:

```
import random
import re
from PIL import Image
from glob import glob
```

```
class FishDataset(Dataset):
    def __init__(self, root_dir, transform=None, target_
    transform=None):
        self.root_dir = os.path.abspath(root_dir)
        self.transform = transform
        self.target_transform = target_transform

        if not self._check_exists():
            raise RuntimeError('Dataset not found.')

        self.images = glob(os.path.join(root_dir, 'fish_image/*
        /*.png'))
        self.masks = [re.sub('fish', 'mask', image) for image in self.
        images]
        print(self.masks[0])
        self.labels = [int(re.search('.*fish_image/fish_(\d+)',
        image).group(1)) for image in self.images]
```

5. Continue to scroll down to the section where we can visualize the image pairs shown in Figure 5-2. We are now using forms, a special feature of Colab that allows us to add input fields directly in the notebook. You can then use those fields to control the output of the cell according to the adjusted variables. In the figure, you can see a sample of the image pairings we will train upon. It's an image of a fish and then a segmented image of the same fish.

VISUALING SAMPLE DATA

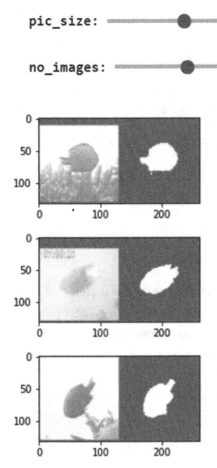

Figure 5-2. *Visualizing fish data input images and masks (segments)*

6. Double-click the image visualization code block to inspect the code, and
    you will see how the magic happens. In the markup you can identify the two
    fields and a special tag on the title { run: "auto" } that tells the cell to
    automatically rerun when any controls are modified. Try it and adjust the image
    size and number of images with the sliders.

```
#@title VISUALING SAMPLE DATA { run: "auto" }
pic_size = 3 #@param {type:"integer"} {type:"slider", min:1, max:30,
step:1}
no_images = 3 #@param {type:"integer"} {type:"slider", min:1, max:32,
step:1}
```

7. Next skip past the other code blocks until you see where the models are created and set up. We are back to creating a single model of type UNet and defining the loss function to binary cross entropy (BCELoss).

```
cuda = True if torch.cuda.is_available() else False
print("Using CUDA" if cuda else "Not using CUDA")

loss_fn = nn.BCELoss()

model = UNet()
if cuda:
  model.cuda()
  loss_fn.cuda()
```

8. Jump down to the training block of code and look inside the batch training loop, as shown here:

```
for batch_idx, (images, masks, _) in enumerate(train_loader):
        images = Variable(images.cuda())
        masks = Variable(masks.cuda())

        optimizer.zero_grad()
        outputs = model(images)
        predicted = outputs.round()
        loss = loss_fn(outputs, masks)
        loss.backward()
        optimizer.step()
```

9. Based on this code, you can see that loss is calculated based on the BCE difference between the outputs out of the model and the input image masks. So, unlike an autoencoder where we train an image to output an image in this case, we train an image to output an image mask, thus requiring our model to learn the mapping from image to mask.

Now, intentionally we missed explaining the UNet model just yet to demonstrate the similarity between this example and an autoencoder. The key difference here is that loss is calculated based on a paired image mask rather than the original image. This allows the model to learn the transformation from image to mask. In the next section, we will look in detail at how a UNet can do this.

Uncovering the Details of a UNet

In simple terms, a UNet is a transformer that allows input images to be translated into outputs not unlike an autoencoder. The key difference between a UNet and an autoencoder is the use of convolutional blocks and how they are joined. Otherwise, the model should feel like an autoencoder but with some key differences.

Of course, to understand this better, we should return to the code from the previous exercise and look in more detail at the UNet model. See Exercise 5-2.

EXERCISE 5-2. EXPLORING THE UNET MODEL

1. Open the GEN_5_UNet.ipynb notebook from the GitHub project site. If you are unsure how, then consult Appendix B.

2. Run the whole notebook by selecting Runtime ➤ Run all if you have not already done so.

3. Scroll down to where the UNet model is defined and double-click the cell to expose the code. We will start looking at the ConvBlock class, which is a helper for wrapping our convolutional layers. That includes initialization of the weights.

```python
class ConvBlock(nn.Module):
    def __init__(self, in_channels,  out_channels):
        super().__init__()
        self.conv = nn.Conv2d(in_channels, out_channels, 3, padding=1)
        init.xavier_uniform(self.conv.weight, gain=np.sqrt(2))
        self.batch_norm = nn.BatchNorm2d(out_channels)
        self.leaky_relu = nn.LeakyReLU(0.01)

    def forward(self, x):
        x = self.conv(x)
        x = self.batch_norm(x)
        x = self.leaky_relu(x)
        return x
```

4. Now move to the start of the UNet class and look at the initialization of the
    network. Notice how we define three down super layers, down1, down2,
    and down3. These consist of ConvBlocks that convert from lower to higher
    channels. They start at 3, then 32, 64, and finish at 128.

    After the middle layer, there are three more super layers, up1, up2, and up3,
    defined to upscale the results back to a single channel. Starting from 128, they
    first upscale to 256, then to 128, and finally to 64. The key difference here
    is that the up layers don't just move a single value up, but it combines them
    from the previous layer. This will become more apparent when we look at the
    forward function.

```
class UNet(nn.Module):
    def __init__(self):
        super().__init__()
        self.down1 = nn.Sequential(
            ConvBlock(3, 32),
            ConvBlock(32, 32)
        )
        self.down2 = nn.Sequential(
            ConvBlock(32, 64),
            ConvBlock(64, 64)
        )
        self.down3 = nn.Sequential(
            ConvBlock(64, 128),
            ConvBlock(128, 128)
        )

        self.middle = ConvBlock(128, 128)

        self.up3 = nn.Sequential(
            ConvBlock(256, 256),
            ConvBlock(256, 64)
        )

        self.up2 = nn.Sequential(
            ConvBlock(128, 128),
            ConvBlock(128, 32)
        )
```

```
            self.up1 = nn.Sequential(
                ConvBlock(64, 64),
                ConvBlock(64, 1)
            )
```

5. Just below in the forward function you can see how all the parts of the UNet are assembled. From this code, you can see how the first three down layers are sequenced together, down1 > down2 > down3 > middle.

```
def forward(self,  x):
    down1 = self.down1(x)
    out = F.max_pool2d(down1, 2)

    down2 = self.down2(out)
    out = F.max_pool2d(down2, 2)

    down3 = self.down3(out)
    out = F.max_pool2d(down3, 2)

    out = self.middle(out)

    out = Upsample(scale_factor=2)(out)
    out = torch.cat([down3, out], 1)
    out = self.up3(out)

    out = Upsample(scale_factor=2)(out)
    out = torch.cat([down2, out], 1)
    out = self.up2(out)

    out = Upsample(scale_factor=2)(out)
    out = torch.cat([down1, out], 1)
    out = self.up1(out)

    out = torch.sigmoid(out)

    return out
```

6. After the middle layer, we move to upsampling by taking the output of the middle and combining it with down3 using `torch.cat` and then passing that through up3. This same process continues so that `middle => out, (out, down3) -> up3 => out, (out, down2)-> up2 => out, (out, down1) -> up1 => out`. Essentially for every pass back through on the way up, we are combining the results with the corresponding down layer. This of course has the effect of reusing the trained features in the down layers while upsampling.

7. At this point, go ahead and alter some of the hyperparameters and see what effect this has on the model output.

8. See if you can alter the architecture of the UNet. Try to modify the number of channels that each super layer uses as input or output. Be sure to account for the merging when passed through the up layers.

While U networks still use convolution to extract features by recycling the input back through the model, we can train better generators that use the same learned feature extraction. Throughout the rest of this chapter, we will explore how the UNet can enhance our generators, especially for image to image generation. In the next section, we look at a new set of GANs that use a UNet for image to image learning.

Translating Images with Pix2Pix

With image segmentation or translation with a UNet, we saw how a model could learn the transformation from one form of image into another. We looked at how a UNet could learn to segment or mask an image based on pairs of training images. The Pix2Pix GAN takes this a step further using adversarial training.

By adding adversarial training via a GAN, we can augment the loss functions we used in a single UNet model and further separate out the interpretation of loss. This not only allows us to do paired base training comparisons for loss but also a second metric using discriminator loss to assure the transformed/generated images more closely resemble their trained pairs. See Exercise 5-3.

EXERCISE 5-3. IMAGE TRANSLATION WITH PIX2PIX

---

1. Open the GEN_5_Pix2Pix.ipynb notebook from the GitHub project site. If you are unsure how, then consult Appendix B.

2. Run the whole notebook by selecting Runtime ➤ Run all if you have not already done so. The first thing to notice is that the Hyperparameter section now provides for alternate training datasets configured using a Colab form. For the image to image examples we provide three datasets. maps represents a map tile pairing of street and satellite images of the same areas. The facades set consists of exterior building shots with pairings of facades displaying blocks representing areas. A third set, cityscapes, is like facades, but instead of buildings, it shows paired annotated views of streets.

    ```
    dataset_name = "maps" #@param ["facades", "cityscapes", "maps"]
    ```

3. We can scroll past major sections of this code since it likely is already familiar to you by now. Move down to the generator/discriminator block definitions. The start of this code shows two UNet classes for helping do upsampling and downsampling.

    ```
    class UNetDown(nn.Module):
        def __init__(self, in_size, out_size, normalize=True,
        dropout=0.0):
            super(UNetDown, self).__init__()
            layers = [nn.Conv2d(in_size, out_size, 4, 2, 1, bias=False)]
            if normalize:
                layers.append(nn.InstanceNorm2d(out_size))
            layers.append(nn.LeakyReLU(0.2))
            if dropout:
                layers.append(nn.Dropout(dropout))
            self.model = nn.Sequential(*layers)

        def forward(self, x):
            return self.model(x)

    class UNetUp(nn.Module):
        def __init__(self, in_size, out_size, dropout=0.0):
            super(UNetUp, self).__init__()
    ```

```
        layers = [
            nn.ConvTranspose2d(in_size, out_size, 4, 2, 1,
            bias=False),
            nn.InstanceNorm2d(out_size),
            nn.ReLU(inplace=True),
        ]
        if dropout:
            layers.append(nn.Dropout(dropout))

        self.model = nn.Sequential(*layers)

    def forward(self, x, skip_input):
        x = self.model(x)
        x = torch.cat((x, skip_input), 1)

        return x
```

4. You may notice ConvTranspose2D being used in the UNetUp class, but that is for the upsampling part. Remember the corresponding down super layer is combined, as shown in the forward function.

5. We can then move down to see how the UNet super layers are used in the GeneratorUNet model.

```
class GeneratorUNet(nn.Module):
    def __init__(self, in_channels=3, out_channels=3):
        super(GeneratorUNet, self).__init__()

        self.down1 = UNetDown(in_channels, 64, normalize=False)
        self.down2 = UNetDown(64, 128)
        self.down3 = UNetDown(128, 256)
        self.down4 = UNetDown(256, 512, dropout=0.5)
        self.down5 = UNetDown(512, 512, dropout=0.5)
        self.down6 = UNetDown(512, 512, dropout=0.5)
        self.down7 = UNetDown(512, 512, dropout=0.5)
        self.down8 = UNetDown(512, 512, normalize=False, dropout=0.5)

        self.up1 = UNetUp(512, 512, dropout=0.5)
        self.up2 = UNetUp(1024, 512, dropout=0.5)
        self.up3 = UNetUp(1024, 512, dropout=0.5)
        self.up4 = UNetUp(1024, 512, dropout=0.5)
```

```
            self.up5 = UNetUp(1024, 256)
            self.up6 = UNetUp(512, 128)
            self.up7 = UNetUp(256, 64)

            self.final = nn.Sequential(
                nn.Upsample(scale_factor=2),
                nn.ZeroPad2d((1, 0, 1, 0)),
                nn.Conv2d(128, out_channels, 4, padding=1),
                nn.Tanh(),

            )
```

6. Notice how we have duplication of several of the UNetDown layers with 512 channels in/out. We can see how this structure is combined in the model's forward function. Notice how down layer d8 is reused as the input into the model and effectively becomes the middle layer.

```
def forward(self, x):
    # U-Net generator with skip connections from encoder to decoder
    d1 = self.down1(x)
    d2 = self.down2(d1)
    d3 = self.down3(d2)
    d4 = self.down4(d3)
    d5 = self.down5(d4)
    d6 = self.down6(d5)
    d7 = self.down7(d6)
    d8 = self.down8(d7)
    u1 = self.up1(d8, d7)
    u2 = self.up2(u1, d6)
    u3 = self.up3(u2, d5)
    u4 = self.up4(u3, d4)
    u5 = self.up5(u4, d3)
    u6 = self.up6(u5, d2)
    u7 = self.up7(u6, d1)

    return self.final(u7)
```

7. Discriminator resembles ones we saw in the past with a key difference. Instead of taking a single three-channel image, this discriminator combines the image pairs into six channels as the input. Notice how this changes the discriminator's forward function.

```
def forward(self, img_A, img_B):
    # Concatenate image and condition image by channels to produce
    input
    img_input = torch.cat((img_A, img_B), 1)
    return self.model(img_input)
```

8. Jump to the bottom of the notebook to the `Training` block. We can see an alteration to the loss calculation in the training loop. Now our discriminator takes the pair of images. This alters the generator loss by needing to account for the loss in reconstructed images but also a pixel-wise comparison. Remember a pixel-wise comparison loss is something we would use in a simple autoencoder. The generator loss is combined with the loss from the discriminator and a scaled pixel-wise loss, where this can be scaled by setting the `hp.lambda_pixel` hyperparameter.

```
# GAN loss
        fake_B = generator(real_A)
        pred_fake = discriminator(fake_B, real_A)
        loss_GAN = criterion_GAN(pred_fake, valid)
        # Pixel-wise loss
        loss_pixel = criterion_pixelwise(fake_B, real_B)

        # Total loss
        loss_G = loss_GAN + hp.lambda_pixel * loss_pixel
```

9. Finally, the discriminator loss is calculated the same way as other GANs, with the only difference being the input of the image pairs.

```
# Real loss
        pred_real = discriminator(real_B, real_A)
        loss_real = criterion_GAN(pred_real, valid)

        # Fake loss
        pred_fake = discriminator(fake_B.detach(), real_A)
        loss_fake = criterion_GAN(pred_fake, fake)

        # Total loss
        loss_D = 0.5 * (loss_real + loss_fake)
```

Training this exercise will output Figure 5-3 using the Maps dataset. From the image you can see that the first row of images represents the original street map image. The image right below is the generated output from the generator, and the image below that is the trained pairing. This is repeated in the next three rows of images.

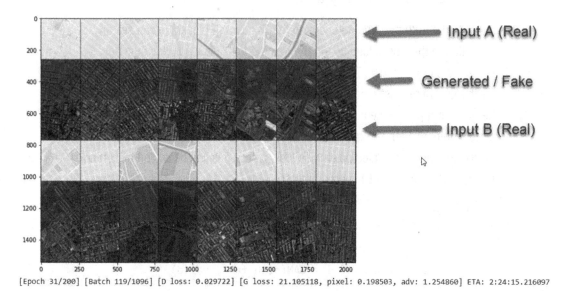

[Epoch 31/200] [Batch 119/1096] [D loss: 0.029722] [G loss: 21.105118, pixel: 0.198503, adv: 1.254860] ETA: 2:24:15.216097

Figure 5-3. *Training output from the Pix2Pix GAN*

If you look at the images, you will notice that the generated fake images have a good amount of detail. This is a result of adding the pixel-wise comparison into the generator loss. Likewise, if we reduced the scale of `lambda_pixel` from 100 to 10/25, then we would see fewer details. Conversely, increasing this value to 1000 would make the output images more closely resemble typical autoencoder output using standard pixel-wise loss.

The Pix2Pix GAN is one of the first image translation models that worked well to produce some interesting results. Pix2Pix and the UNet are used extensively in various imaging applications requiring segmentation and/or transformation. In the next section, we expand on this image translation model by doubling up with the DualGAN.

Seeing Double with the DualGAN

It stands to reason that if we can train a GAN to perform image to image translation going one way, from a street image to satellite image, for instance, then we could likewise reverse the process. We could then also perhaps account for additional metrics of loss by understanding how each pair of generator/discriminator trains.

With the DualGAN, we perform both image to image translations with a pair of generators and discriminators using the image pairings. This allows us to take a combined loss of training images from one domain to the other and back again. A process known as *cycle consistency loss*.

Cycle consistency loss is a method we will explore heavily in the next chapter with more examples. Until then, Figure 5-4 shows how cycle loss is calculated based on the image pairings. To calculate cycle loss, the first set of generated images is fed back into the opposing generator in order to generate new X/Y images based on fake images. This conversion and back again is known as *cycle consistency*.

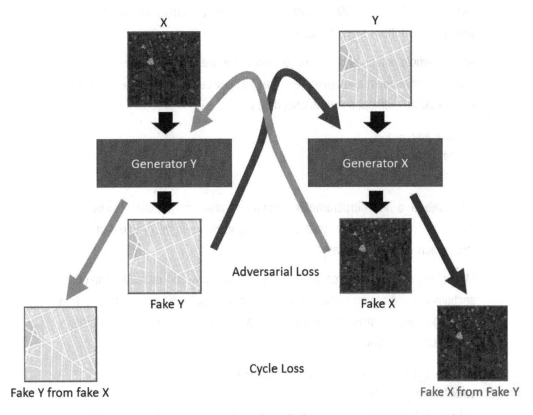

Figure 5-4. *Calculating adversarial and cycle loss*

By using cycle consistency, both generators are trained at the same time using alternate inputs of what the other generator generated as real. This allows for both generators now to be tightly coupled to the cycle loss.

For the discriminator, we also enhance the method of loss calculation using a Wasserstein gradient penalty calculation. If you recall, a Wasserstein GAN (WGAN) calculated the relative distance between two sets of distributed data. The addition of a gradient penalty is a method to soften the WGAN loss function.

When we put all this together, we get our first instance of a double GAN, the DualGAN, as we will demonstrate in Exercise 5-4. This GAN uses the combination of two generators and two discriminators to work in tandem. This means the calculation of loss for the paired discriminator and generator is in turn combined. We will see how this all comes together in our next exercise.

EXERCISE 5-4. SEEING DOUBLE WITH THE DUALGAN

1. Open the GEN_5_DualGAN.ipynb notebook from the GitHub project site. If you are unsure how, then consult Appendix B.

2. Run the whole notebook by selecting Runtime ➤ Run all if you have not already done so. Scroll down and open the Hyperparameters section; then see the new variables defined, as summarized here:

```
lambda_adv = 1,
lambda_cycle = 10,
lambda_gp = 10
```

3. The lambda_adv hyperparameter is the adversarial loss modifier. lambda_cycle is used to scale the cycle loss, and lambda_gp is the gradient penalty loss, which we will cover later.

4. Most of the code we have covered a couple times now, so just review the other sections of code as you scroll down to the training block. Be sure to look at how the models are constructed. We will first look at how the loss for the generators is calculated, as shown here:

```
# Translate images to opposite domain
fake_A = G_BA(imgs_B)
fake_B = G_AB(imgs_A)
```

```
# Reconstruct images
recov_A = G_BA(fake_B)
recov_B = G_AB(fake_A)

# Adversarial loss
G_adv = -torch.mean(D_A(fake_A)) - torch.mean(D_B(fake_B))
# Cycle loss
G_cycle = cycle_loss(recov_A, imgs_A) + cycle_loss(recov_B, imgs_B)
# Total loss
G_loss = hp.lambda_adv * G_adv + hp.lambda_cycle * G_cycle
```

5. We have two generators, one for generating images from domain A to B and another going from B to A named G_AB and G_BA, respectively. Notice how we pass the fake/generated images back into the opposing generators to create recov_A and recov_B outputs. The reconverted images are used to calculate the cycle consistency loss. After the various images are generated, then we first calculate adversarial loss by passing in the fakes to the respective discriminator D_A or D_B. Next, cycle loss is calculated from comparisons between the reconverted images and originals. Thus, we determine how much error is in a full translation from generator A > B and B > A. Finally, the total G_loss is calculating by summing up the values and applying the lambda scaling hyperparameters.

6. Next look at the discriminator section and look at the loss calculation shown here:

```
# Compute gradient penalty for improved wasserstein training
gp_A = compute_gradient_penalty(D_A, imgs_A.data, fake_A.data)
# Adversarial loss
D_A_loss = -torch.mean(D_A(imgs_A)) + torch.mean(D_A(fake_A)) +
hp.lambda_gp * gp_A

# Compute gradient penalty for improved wasserstein training
gp_B = compute_gradient_penalty(D_B, imgs_B.data, fake_B.data)
# Adversarial loss
D_B_loss = -torch.mean(D_B(imgs_B)) + torch.mean(D_B(fake_B)) +
hp.lambda_gp * gp_B

# Total loss
D_loss = D_A_loss + D_B_loss
```

```
D_loss.backward()
optimizer_D_A.step()
optimizer_D_B.step()
```

7. For the discriminator loss, we are using the special function `compute_gradient_penalty` that we will get too shortly. This function takes the discriminator (A or B) and the real and fake images for the respective domain (A or B). From this we calculate two forms of adversarial loss; `D_A_loss`/`D_B_loss` combines normal adversarial loss plus the previous gradient penalty loss again scaled by `lambda_gp`. Finally, total loss is combined from the two domain A/B losses. Notice at the end of the code how we use two different optimizers to train the weights in the discriminators. This contrasts with the generators, which use only a single optimizer.

8. Our last step will be to move back up to the `compute_gradient_penalty` function, as shown here:

```
def compute_gradient_penalty(D, real_samples, fake_samples):
    alpha = FloatTensor(np.random.random((real_samples.size(0),
    1, 1, 1)))

    interpolates = (alpha * real_samples + ((1 - alpha) * fake_
    samples)).requires_grad_(True)
    validity = D(interpolates)
    fake = Variable(FloatTensor(np.ones(validity.shape)), requires_
    grad=False)
    # Get gradient w.r.t. interpolates
    gradients = autograd.grad(
        outputs=validity,
        inputs=interpolates,
        grad_outputs=fake,
        create_graph=True,
        retain_graph=True,
        only_inputs=True,
    )[0]
    gradients = gradients.view(gradients.size(0), -1)
    gradient_penalty = ((gradients.norm(2, dim=1) - 1) ** 2).mean()
    return gradient_penalty
```

9. This function is based on the Wasserstein GP loss from Chapter 4. Recall that this converts the Wasserstein earth mover's method of measuring differences in distributions using a gradient penalty rather than a clipping function.

As shown in Figure 5-5, this GAN is quite efficient at learning image to image translation across both domains and back. This results in the figure showing the domain translation from A to B and back again. After just a few epochs, the double GANs are making short work of learning the domain to domain translation.

From image to image paired translation, we can start to move on to consider other possibilities of image translation models. In the next section, we consider our first model that assumes not every image has one correct translation.

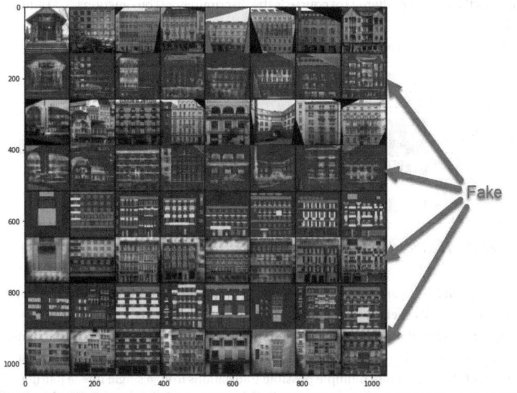

[Epoch 199/200] [Batch 60/64] [D loss: -0.271755] [G loss: 0.689006, cycle: 0.096770] ETA: 0:00:00.626353

Figure 5-5. *Training output of DualGAN for image to image translation*

Riding the Latent Space on the BicycleGAN

A major downfall in image paired translations is the assumption that each image has only one correct translation. This is often not the case, and one image or phrase in a language could have multiple correct meanings in the other domain.

The BicycleGAN introduced the assumption that every input image had multiple correct translations and not just one. However, to perform this feat, this form of GAN uses an old friend, the variational autoencoder, to better map across the learned distribution.

Now, instead of paired generators and discriminators, this model uses a generator and VAE encoder, as shown in Figure 5-6. From the figure, A and B denote the different image domains, where $\hat{B}$ represents the output of A. The G denotes the generator, and E denotes the encoder, where D of course is the discriminator. The functions $N(z)$ and $Q(z|B)$ are sampling functions where the output is passed to the KL block. KL stands for Kullback-Leibler and is another way to measure differences in distribution outputs.

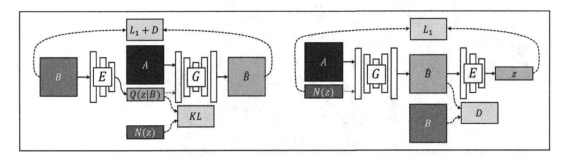

Figure 5-6. *The BicycleGAN architecture*

The *L1* and *L1+D* denotes the cycle loss or the amount of loss from translating to and back again. We will cover other examples of cycle loss in Chapter 6.

Figure 5-6 should provide the basic intuition behind how the BicycleGAN functions, but of course we want to dig into the code and see it run as well. In Exercise 5-5, we use the BicycleGAN to generate multiple possible translations from a single image pairing.

EXERCISE 5-5. RIDING THE LATENT SPACE

1. Open the GEN_5_BicycleGAN.ipynb notebook from the GitHub project site. If you are unsure how, then consult Appendix B. Run the whole notebook by selecting Runtime ➤ Run all if you have not already done so. Scroll down and open the Hyperparameters section to see the new variables defined, as summarized here:

```
lambda_pixel=10,
lambda_latent=.5,
lambda_kl=.01
```

2. These hyperparameters are all used to scale the various loss outputs. Where pixel denotes pixel-wise loss, latent denotes the latent space encoding loss from the encoder and kl the Kullback-Leibler loss from the learned distributions.

3. Scroll down to the models' section and specifically to the Encoder model, as shown here:

```
class Encoder(nn.Module):
  def __init__(self, latent_dim, input_shape):
    super(Encoder, self).__init__()
    resnet18_model = resnet18(pretrained=False)
    self.feature_extractor = nn.Sequential(*list(resnet18_model.
    children())[:-3])
    self.pooling = nn.AvgPool2d(kernel_size=8, stride=8, padding=0)
    # Output is mu and log(var) for reparameterization trick used in VAEs
    self.fc_mu = nn.Linear(256, latent_dim)
    self.fc_logvar = nn.Linear(256, latent_dim)

  def forward(self, img):
    out = self.feature_extractor(img)
    out = self.pooling(out)
    out = out.view(out.size(0), -1)
    mu = self.fc_mu(out)
    logvar = self.fc_logvar(out)
    return mu, logvar
```

4. There are a couple key things to note about the `Encoder` class. First this encoder is using an existing model called `resnet18` as a feature extractor. We will learn more about the ResNet model in Chapter 6 as well as how we can reuse pretrained models. Using pretrained models is known as *transfer learning*.

5. Next, we will look at a summary of how the models and optimizers are created here:

```
generator = Generator(hp.latent_dim, input_shape)
encoder = Encoder(hp.latent_dim, input_shape)
D_VAE = MultiDiscriminator(input_shape)
D_LR = MultiDiscriminator(input_shape)

optimizer_E = torch.optim.Adam(encoder.parameters(), lr=hp.lr,
betas=(hp.b1, hp.b2))
optimizer_G = torch.optim.Adam(generator.parameters(), lr=hp.lr,
betas=(hp.b1, hp.b2))
optimizer_D_VAE = torch.optim.Adam(D_VAE.parameters(), lr=hp.lr,
betas=(hp.b1, hp.b2))
optimizer_D_LR = torch.optim.Adam(D_LR.parameters(), lr=hp.lr,
betas=(hp.b1, hp.b2))
```

6. We can see four models constructed, each with their own optimizer. Note how we differentiate the discriminator loss models, as D_VAE measures encoder accuracy and D_LR measures generator accuracy.

7. Finally, the training code should mostly be familiar aside from the model configuration. Consulting Figure 5-6 should help explain the code. We will focus on one key section, as shown here:

```
mu, logvar = encoder(real_B)
encoded_z = reparameterization(mu, logvar)
fake_B = generator(real_A, encoded_z)

# Pixelwise loss of translated image by VAE
loss_pixel = mae_loss(fake_B, real_B)
# Kullback-Leibler divergence of encoded B
loss_kl = 0.5 * torch.sum(torch.exp(logvar) + mu ** 2 - logvar - 1)
# Adversarial loss
loss_VAE_GAN = D_VAE.compute_loss(fake_B, valid)
```

8. The tricky or confusing part about this code is the output from the encoder model; mu and `logvar` (variance) are used as inputs into the `reparameterization` function. This function just converts the values and resamples from the space. Recall how this is done in the standard VAE. This encoded value is then passed into the generator along with a real image. We then measure loss using pixel-wise comparison, KL, and the discriminator loss. Be sure to give the rest of the loss calculation code a further review.

9. At the bottom of the notebook, you will see the training output, as shown in Figure 5-7. In the output you can see how for each input image several variations are output. These variations are controlled by using the VAE encoder to learn the domain distribution used in the model.

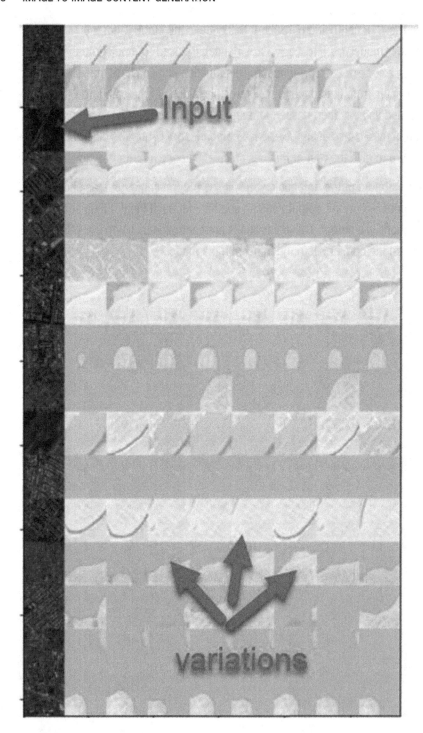

Figure 5-7. *Example output from training the BicycleGAN*

The concept behind the BicycleGAN is paramount to how we want to proceed when thinking about translation pairings. However, in terms of accuracy, this model lacks the ability to map across domains well, as demonstrated in Figure 5-7. Going a step further, we will next consider how to translate across domains without pairings.

Discovering Domains with the DiscoGAN

The ability to translate images or languages is often done with paired datasets. For language, this will often be comparable phrases in both languages, for example. In the case of our previous images, pairings of the same image were translated in some form.

Now, of course, getting accurate translation across languages using unpaired phrases is not likely. However, what can be learned across these types of translation tasks using unpairings is the domain essence or style.

The DiscoGAN is designed to capture and allow for a translation of style across unpaired images. It does this by combining the adversarial, pixel, and cycle loss across both generators. The effect allows for the translation of horses to zebras or apples to oranges, as we will see in Exercise 5-6.

EXERCISE 5-6. DISCOVERING DOMAINS WITH A DISCOGAN

1. Open the GEN_5_DiscoGAN.ipynb notebook from the GitHub project site. If you are unsure how, then consult Appendix B.

2. This GAN provides for a number example datasets that we will use for this example and several more in later chapters. Figure 5-8 shows a summary of the available datasets and what the matching images look like. Be sure to select your desired dataset and then run the whole notebook. Select Run ➤ Run all from the menu.

Apples 2 Oranges

Monet 2 Photo

Horse 2 Zebra

Summer 2 Winter Yosemite

VanGogh 2 Photo

Figure 5-8. *Examples of unpaired image to image datasets*

3. Feel free to explore the various datasets and get comfortable with how they look through training. This will be good practice for when we explore other example exercises that use these sets again.

4. Most of the code in this sample should be review by now. So, we will focus on just the creation of the models and optimizers, as shown here:

```
G_AB = GeneratorUNet(input_shape)
G_BA = GeneratorUNet(input_shape)
D_A = Discriminator(input_shape)
```

```
D_B = Discriminator(input_shape)

optimizer_G = torch.optim.Adam(
    itertools.chain(G_AB.parameters(), G_BA.parameters()), lr=hp.lr,
    betas=(hp.b1, hp.b2)
)
optimizer_D_A = torch.optim.Adam(D_A.parameters(), lr=hp.lr,
betas=(hp.b1, hp.b2))
optimizer_D_B = torch.optim.Adam(D_B.parameters(), lr=hp.lr,
betas=(hp.b1, hp.b2))
```

5. Again, we have two generators and discriminators in the GAN. While both generators share an optimizer, the discriminators use their own.

6. Jump down to the training code and specifically the loss calculations. We will focus on how the cycle loss is calculated in the generators, as shown here:

```
fake_B = G_AB(real_A)
loss_GAN_AB = adversarial_loss(D_B(fake_B), valid)
fake_A = G_BA(real_B)
loss_GAN_BA = adversarial_loss(D_A(fake_A), valid)

loss_GAN = (loss_GAN_AB + loss_GAN_BA) / 2

# Pixelwise translation loss
loss_pixelwise = (pixelwise_loss(fake_A, real_A) + pixelwise_
loss(fake_B, real_B)) / 2

loss_cycle_A = cycle_loss(G_BA(fake_B), real_A)
loss_cycle_B = cycle_loss(G_AB(fake_A), real_B)
loss_cycle = (loss_cycle_A + loss_cycle_B) / 2

# Total loss
loss_G = loss_GAN + loss_cycle + loss_pixelwise
```

7. Adversarial loss (loss_GAN) is first calculated and then pixel-wise and finally cycle loss. All three are combined into the total generator loss (loss_G).

8. Let the model train on the various datasets. Be sure to reset the models after each time you switch the datasets. Otherwise, you will experience strange results.

9. You may also choose to train multiple datasets in succession, thus allowing the GAN to learn multiple domain styles and combine them. Figure 5-9 shows an example of such a training that was first trained on the Monet 2 Photo set and then switches to the VanGogh 2 Photo set.

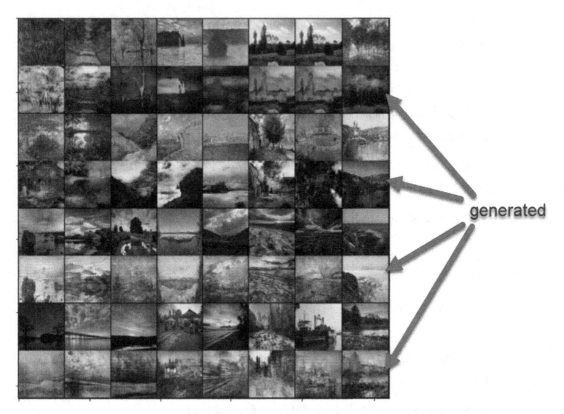

generated

Figure 5-9. *Example of cross-domain training, from VanGogh to Monet*

The output from this exercise is interesting because it shows how the domain-style translation is put into effect. In some cases, the art and photo look believable after a short amount of training. As you can see, the DiscoGAN is an effective solution for translating domains/styles across images.

Be sure to explore and play with the previous exercise using multiple variations or different data. Visualizing how this model trains across domains and cross domains can be enlightening, especially when you try to tune some of the hyperparameters in the model.

Conclusion

In this chapter, we looked at increasing the bandwidth of our image generation to an image translation/generation model. We first explored this concept with the encoding structure of the UNet for performing image segmentation, where we looked at segmenting images of fish. Then we employed the UNet model in an image-to-image translation with Pix2Pix and later upgraded to the DualGAN, a GAN using two generators and discriminators.

From understanding image to image translation, we moved on the exploring other forms of image or domain translation. First we looked at the BicycleGAN, which used the assumption that for each input multiple translations exist. Then we moved on to the DiscoGAN for performing unpaired image-style domain transfer.

The techniques we have used in this chapter have extended to many areas that typically use images for analysis. As an example in the medical field, image to image and segmentation GANs provide excellent support for medical diagnostics, even allowing these GANs to surpass medical experts in their field by identifying diagnostic markers for everything from breast cancer to heart problems.

In the next chapter, we will continue our exploration of domain translation with unpaired images and how we can further control such output variations.

CHAPTER 6

Residual Network GANs

Generative adversarial networks and adversarial training are truly limitless in concept but often fall short in execution and implementation. As we have seen throughout this book, the failures often reside in the generator. And, as we have learned, the key to a good GAN is a good generator.

The discriminator in a GAN is simply a classifier and often a simple binary one at that. In fact, it is a common practice now to build a GAN to train a discriminator/classifier. A GAN will be trained on a set of data for binary classification, and after training, the generator is discarded. Not only does this provide an accurate classifier but one that is also robust as well.

Through adversarial learning, the classifier can learn better approximations across a wider training space provided by the generator. The generator can fill in those latent spaces by trying to trick the discriminator/classifier to better learn and approximate to. Thus, a classifier trained on real data also learns to spot approximated fakes of the same data.

As it turns out, building a good discriminator is considerably easier than creating a good generator, and this is due to the learning problem. Generators require a finesse in understanding features and the target distribution so that they can reproduce said features, while discriminators only need to classify the differences or distance between features.

We saw early on how convolutional neural networks can help extract those features. But we saw those features would often be exposed as patches or blobs when regenerated. Then we looked at the UNet for being able to translate learned features back to generated output. As we saw in Chapter 5, this can produce better generated output.

In this chapter, we take the next step or evolution of feature extracting mechanisms, the residual block and residual network (ResNet). We will break down and understand how the ResNet and residual blocks work for classification. We learn how these models are able to bypass the depth layer problem we encounter with deep CNN networks.

© Micheal Lanham 2021
M. Lanham, *Generating a New Reality*, https://doi.org/10.1007/978-1-4842-7092-9_6

As the models we develop continue to use larger more complex feature extractors in various forms, we will move to understanding how to reuse existing previously trained models. Reusing previously trained standardized models is called *transfer learning*, and we will explore a simple example using a pretrained model for classification.

We will also put all this new knowledge together and look at three GAN variations that use ResNets to build better generators. The first of these will be the CycleGAN and extended unpaired image to image generator. The second will be a GAN dedicated to learning and conditionally creating faces from looking at celebrity faces.

Finally, we will finish the chapter looking at the super resolution GAN (SRGAN), a GAN that combines the best of multiple models from a ResNet to a pretrained robust feature extractor. This is where we will use the SRGAN to improve resolution on some celebrity faces.

As we progress through the book, the examples get far more complicated, so to save space, only important snippets are shown. It will likely be more helpful to have the code open while you are working through the chapter. Here again is what we will cover in this chapter:

- Understanding residual networks

- Cycling again with CycleGAN

- Creating faces with StarGAN

- Using the best with transfer learning

- Increasing resolution with SRGAN

We will continue to build on our knowledge from previous chapters. If you need to, be sure to review the content on convolutional networks and feature extraction/ generation. Residual networks are big consumers of convolutional layers, as we will see in the next section.

Understanding Residual Networks

As we saw in previous chapters and exercises, we often need to increase the depth of our networks for better learning feature extraction. However, as we have also learned, the deeper those network layers became, the more problems arose such as exploding or vanishing gradients that are often caused by feature over feature extraction.

Feature over feature extraction happens when convolutional network layers become too deep and, even though they may be normalized, still create difficult to map latent surfaces. The effect is the model over-emphasizes some features and ignores fewer less common ones.

We have seen plenty examples of feature over feature extraction in many GANs that use multiple convolutional layers. This can be identified as the bright areas in regenerated images.

Residual networks and blocks try to overcome this feature over feature effect by allowing some layers to be skipped, thus allowing the model to ignore layers that may be overemphasizing features. ResNets do this by passing residual outputs from the top of the block to the bottom after a convolutional process.

Figure 6-1 shows a single residual block. In the figure, the inputs x come in and enter the first weight layer and bypass it to the + node, the bottom. $F(x)$ represents the convolutional function applied to x using the weight layers. At the bottom of the block, you can see that the output is $F(x) + x$, where x is the residuals from the start of the block.

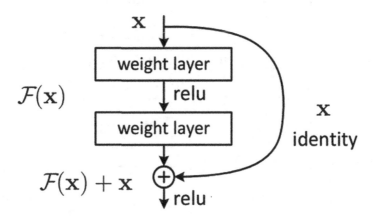

Figure 6-1. *A single residual block*

By passing the residual inputs from the top of the block to the bottom, we effectively reduce and normalize the dependency on the weight layers. This in turn reduces exploding/vanishing gradients caused by feature over feature extraction. Another benefit of this is the ability to increase network depth significantly.

Figure 6-2 shows a comparison of a 19-layer network, the reference VGG19, and ResNet152, which is a residual network that uses 152 layers. Within the ResNet model, you can see the skip connections between the residual convolution blocks. Consider how deep ResNet152 is compared to the simple VGG19.

Visual Geometry Group (VGG) from Oxford University developed a set of standardized models that used convolution. In VGG19 there are 19 layers, 16 convolution, and 3 linear flat layers.

We can see how a residual block within a residual network works for a simple classification problem by again looking at the fashion MNIST dataset. This time we are going to build a simple model with a residual network to learn how to classify fashion. Jump on to your computer and get ready to explore Exercise 6-1.

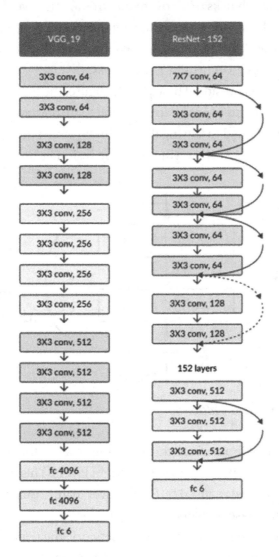

Figure 6-2. *Comparison of regular convolutional network and ResNet*

EXERCISE 6-1. RESIDUAL NETWORK CLASSIFIER

1. Open the GEN_6_ResNet_classifier.ipynb notebook from the GitHub project site. If you are unsure how, then consult Appendix B.

2. Run the whole notebook by selecting Runtime ➤ Run all. The bulk of this notebooks code should be familiar. Therefore, we will focus only on the new or important sections.

3. You can scroll down to the RESNET model block and look at the first two helper functions. The first of these functions, preprocess, is for swapping and setting up the tensors. The second is the conv function that is a wrapper around creating a convolutional layer.

```
def preprocess(x):
    return x.view(-1, 1, 28, 28)

def conv(in_size, out_size, pad=1):
    return nn.Conv2d(in_size, out_size, kernel_size=3, stride=2,
    padding=pad)
```

4. Just below that is the definition for the ResidualBlock class. Remember, a single residual block allows for the input residuals to bypass the training layers and then get added in at the bottom with the skip connection.

```
class ResBlock(nn.Module):
    def __init__(self, in_size:int, hidden_size:int, out_size:int,
    pad:int):
        super().__init__()
        self.conv1 = conv(in_size, hidden_size, pad)
        self.conv2 = conv(hidden_size, out_size, pad)
        self.batchnorm1 = nn.BatchNorm2d(hidden_size)
        self.batchnorm2 = nn.BatchNorm2d(out_size)

    def convblock(self, x):
        x = nn.functional.relu(self.batchnorm1(self.conv1(x)))
        x = nn.functional.relu(self.batchnorm2(self.conv2(x)))
        return x

    def forward(self, x): return x + self.convblock(x) # skip
    connection
```

171

5. Below that is where the `ResidualNetwork` class is constructed. Notice how there are two `ResBlocks` residual blocks connected in sequence and the output is passed through a batch normalization and max pooling at the end/output.

```
class ResNet(nn.Module):
    def __init__(self, n_classes=10):
        super().__init__()
        self.res1 = ResBlock(1, 8, 16, 15)
        self.res2 = ResBlock(16, 32, 16, 15)
        self.conv = conv(16, n_classes)
        self.batchnorm = nn.BatchNorm2d(n_classes)
        self.maxpool = nn.AdaptiveMaxPool2d(1)

    def forward(self, x):
        x = preprocess(x)
        x = self.res1(x)
        x = self.res2(x)
        x = self.maxpool(self.batchnorm(self.conv(x)))
        return x.view(x.size(0), -1)
```

6. The next thing to focus on are various snippets or lines of important code, shown next. The first line shows the loss function we are using. Cross-entropy loss is standard for classification problems. After that there is a function called `accuracy`. The `accuracy` function returns a percentage for how accurately a classification matched a set of outputs. It works by taking the maximum predicted values and finding the index using `argmax`. It then compares the correct outputs with what is labeled, returning a value between 0.0 and 1.0, or 0 to 100 percent.

```
loss_fn = nn.CrossEntropyLoss()

def accuracy(pred, labels):
    preds = torch.argmax(pred, dim=1)
    return (preds == labels).float().mean()
```

7. Let the model train until completion, and notice the output as the model trains. We should see loss decreasing for both the training and test sets, and we should see the accuracy increase for both sets. If the model is training well, both curves in loss and accuracy should follow each other. Loss or accuracy that diverges between test and training is a result of over- or underfitting.

8. Numbers aside, it is often helpful to see actual results, so the last block of code on the notebook provides actual visual confirmation. Inside this block we sample a batch of images/labels from the testloader and then pass them into the model to output predictions, `preds`. We then need to convert those predictions back to labels/indices with `torch.max`. After that, the model accuracy is printed, multiplied by 100, and the images are plotted with the labels.

```
dataiter = iter(testloader)
images, labels = dataiter.next()

preds = model(images.cuda())
values, indices = torch.max(preds, 1)
print(f"accuracy {accuracy(preds, labels.cuda()).item()*100}%")
plot_images(images.cpu().numpy(), indices.cpu().numpy(), 25)
```

9. Figure 6-3 shows a set of images with predicted labels. While the accuracy is at over 92 percent, we can see in the output image that no images are incorrectly labeled.

One thing you ideally noticed when running that previous exercise is how quickly this model trained. This model is initially set to train with only five epochs and in most cases can quickly get up to 90 percent accuracy. If you train this model longer, it should easily exceed 90 percent accuracy for this dataset.

As you can see, residual networks provide increased stability to deeper networks and can allow for an increase in training performance. In turn, they allow for deeper networks that can learn feature sets that don't suffer from feature over feature extraction.

With another new powerful addition to our toolbelt, we can move back to looking at how this improves on GAN variations. We will start back again to looking at unpaired image to image translation with the CycleGAN next.

accuracy 92.1875%

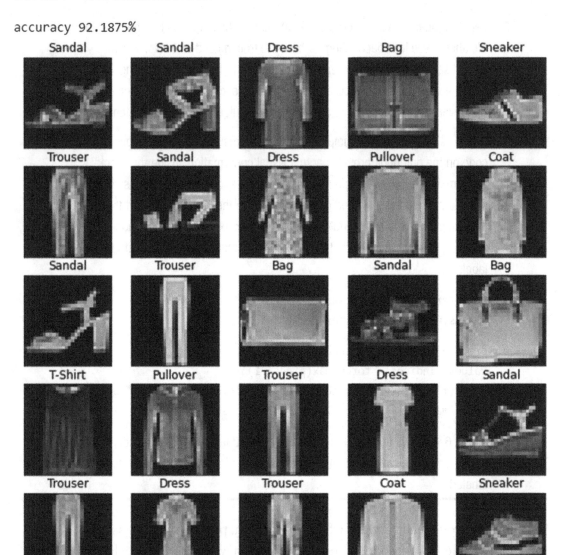

Figure 6-3. *Predicted accuracy and labeled output*

Cycling Again with CycleGAN

The cycle or cycle consistency GAN's primary loss mechanism is cycle loss. Recall that we previously covered cycle loss in Chapter 5. Cycle consistency loss is measured by the translation of an item into something and then translated back into the original, where the measured loss is taken by how well the original compares with the double translation output.

While the CycleGAN is named after its primary method of loss determination, it again combines other methods of loss as assurances. This variation of a GAN uses three forms of loss on the generator: the adversarial or standard GAN loss, the cycle loss, and an identity loss.

Identity loss measures the color variation or temperature change over an image. Without the addition of identity loss, the generated output may take on different hues much different than our desired output. A CycleGAN doesn't necessarily need to include the identity loss, but it makes the output more uniform.

Since we have already covered a variation of this GAN, the BicycleGAN and DiscoGAN, we can quickly jump into the exercise. Exercise 6-2 will be similar to the DiscoGAN exercise we looked at in Chapter 5. There may be a few variations like identity loss that we will cover further as we get into the exercise.

EXERCISE 6-2. CYCLING AND IDENTITY WITH CYCLEGAN

1. Open the GEN_6_CycleGAN.ipynb notebook from the GitHub project site. If you are unsure how, then consult Appendix B.

2. The bulk of this notebook's code should be familiar, but we will again cover the new or different hyperparameters shown. The n_residual_blocks hyperparameter allows for us to control the number of residual blocks used in the generator, where lambda_cyc and lambda_id control the scale on the cycle and identity loss, respectively.

    ```
    n_residual_blocks=9,
    lambda_cyc=10.0,
    lambda_id=5.0
    ```

3. You will also note that this notebook allows you to choose between the five unpaired image to image datasets: Monet to Photo, VanGogh to Photo, Apples to Oranges, Summer to Winter in Yosemite, and Horses to Zebras. Select a dataset that you find interesting and then run the whole notebook by selecting Runtime ➤ Run all.

4. You may notice that the GeneratorResNet class is comprised of
    downsampling layers joined to several residual blocks, set by the
    hyperparameter, that output to upsample layers for the generated output. An
    abridged version of the init function shows how the model architecture is
    assembled.

```python
def __init__(self, input_shape, num_residual_blocks):
    super(GeneratorResNet, self).__init__()

    channels = input_shape[0]
    out_features = 64
    model = [
        nn.ReflectionPad2d(channels),
        nn.Conv2d(channels, out_features, 7),
        nn.InstanceNorm2d(out_features),
        nn.ReLU(inplace=True),
    ]
    in_features = out_features

    # Downsampling
    for _ in range(2):
        out_features *= 2
        model += [
            nn.Conv2d(in_features, out_features, 3, stride=2,
            padding=1),
            nn.InstanceNorm2d(out_features),
            nn.ReLU(inplace=True),
        ]
        in_features = out_features

    # Residual blocks
    for _ in range(num_residual_blocks):
        model += [ResidualBlock(out_features)]

    # Upsampling
    for _ in range(2):
        out_features //= 2
        model += [
            nn.Upsample(scale_factor=2),
```

```
            nn.Conv2d(in_features, out_features, 3, stride=1,
            padding=1),
            nn.InstanceNorm2d(out_features),
            nn.ReLU(inplace=True),
        ]
        in_features = out_features

    # Output layer
    model += [n...
```

5. Next, we will look at the loss, model creation, and optimizer functions.
    Notice how the generator shares an optimizer, while the discriminator uses
    independent optimizers.

```
# Loss functions
criterion_GAN = torch.nn.MSELoss()
criterion_cycle = torch.nn.L1Loss()
criterion_identity = torch.nn.L1Loss()

input_shape = (hp.channels, hp.img_size, hp.img_size)

# Initialize generator and discriminator
G_AB = GeneratorResNet(input_shape, hp.n_residual_blocks)
G_BA = GeneratorResNet(input_shape, hp.n_residual_blocks)
D_A = Discriminator(input_shape)
D_B = Discriminator(input_shape)

optimizer_G = torch.optim.Adam(
    itertools.chain(G_AB.parameters(), G_BA.parameters()), lr=hp.lr,
    betas=(hp.b1, hp.b2)
)
optimizer_D_A = torch.optim.Adam(D_A.parameters(), lr=hp.lr,
betas=(hp.b1, hp.b2))
optimizer_D_B = torch.optim.Adam(D_B.parameters(), lr=hp.lr,
betas=(hp.b1, hp.b2))
```

6. Right after the optimizers are created, we add a new tool called a *scheduler*.
    Schedulers are used to modify hyperparameters during training. In this case,
    we used one to modify the learning rate during training. For this model, we
    decrease the learning rate over time, thus allowing our model to start out
    making big changes and then over time gradually make small tweaks.

```
# Learning rate update schedulers
lr_scheduler_G = torch.optim.lr_scheduler.LambdaLR(
    optimizer_G, lr_lambda=LambdaLR(hp.n_epochs, hp.epoch, hp.decay_
    epoch).step
)
lr_scheduler_D_A = torch.optim.lr_scheduler.LambdaLR(
    optimizer_D_A, lr_lambda=LambdaLR(hp.n_epochs, hp.epoch, hp.decay_
    epoch).step
)
lr_scheduler_D_B = torch.optim.lr_scheduler.LambdaLR(
    optimizer_D_B, lr_lambda=LambdaLR(hp.n_epochs, hp.epoch, hp.decay_
    epoch).step
)
```

7. Jump down to the training code, where we will first look at how the identity loss is calculated.

```
loss_id_A = criterion_identity(G_BA(real_A), real_A)
loss_id_B = criterion_identity(G_AB(real_B), real_B)

loss_identity = (loss_id_A + loss_id_B) / 2
```

8. Next the GAN or adversarial loss is like many previous examples:

```
fake_B = G_AB(real_A)
loss_GAN_AB = criterion_GAN(D_B(fake_B), valid)
fake_A = G_BA(real_B)
loss_GAN_BA = criterion_GAN(D_A(fake_A), valid)

loss_GAN = (loss_GAN_AB + loss_GAN_BA) / 2
```

9. Finally, the cycle loss is like what we did in Chapter 4, and then we calculate the final loss using the lambda scalings:

```
recov_A = G_BA(fake_B)
loss_cycle_A = criterion_cycle(recov_A, real_A)
recov_B = G_AB(fake_A)
loss_cycle_B = criterion_cycle(recov_B, real_B)
```

```
loss_cycle = (loss_cycle_A + loss_cycle_B) / 2

# Total loss
loss_G = loss_GAN + hp.lambda_cyc * loss_cycle + hp.lambda_id * loss_
identity
```

While running this example on your chosen dataset, you may notice some of those bright patches appearing. This is being caused by feature over feature extraction, and it should quickly go away. If you find it persists, you can increase the hyperparameter n_residual_blocks to increase the number of residual blocks in the generator.

Increasing the number of blocks in the residual generator will reduce the feature over feature extraction issues but will take longer to train. The longer training time is a result of the increased number of weights/parameters in the network.

Figure 6-4 shows a subset of the results of training the CycleGAN on Horse to Zebras over 200 epochs using a residual generator with 9 residual blocks. The results are not optimal, but as you can see, the generator does a good job. We could, of course, improve on this training example with tweaking and hyperparameter tuning.

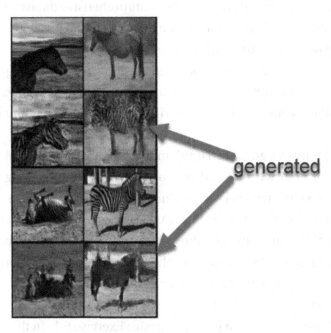

Figure 6-4. *Sample training output from CycleGAN*

One thing that can make the translation more natural is adding residual blocks. You can go back and try adding residual blocks using the hyperparameter and watch the effect on training. You may also want to switch datasets and retrain to see the effect swapping the data domain has.

ResNet provides a useful set of tools to the application of image to image translation by learning to replicate features. It can also be useful in other applications of classification as we saw and other forms of image translation. We will look at a new form of conditioned image translation in the next section.

Creating Faces with StarGAN

Up to this point we have looked at unconditional image to image translation using paired or unpaired training sets. As we learned through our progression of GANs, one useful feature is the ability to condition output. Not only does this provide us with control of the output, but it also increases the training performance of the model.

One such GAN devised to introduce conditioning was the StarGAN. It was so named because the GAN is conditionally trained on a comprehensive dataset of celebrity faces that have been annotated with attributes. The dataset's name is CelebA, and it has become the standard for face training datasets.

CelebA by itself comes in a few flavors. The version we will be using has the faces centered in the frame. Along with this, we will be using an annotated text file that attempts to describe the images with 30+ attributes from black to blond hair, male, and young.

StarGAN itself is just an advanced implementation of a conditioned GAN such as the cGAN/CGAN or cDCGAN we reviewed previously. The incredible thing about StarGAN is how well it uses residual networks to produce some incredible output.

Figure 6-5 shows the output of the StarGAN notebook we will look at in Exercise 6-3. The GAN was conditionally trained on these classes: black hair, blond hair, brown hair, male, and young. Based on the output from the figure, it looks like it does a generally good job aside from perhaps the young attribute.

Since we reviewed a conditioned GAN previously and already covered image to image translation, we can move on to working with Exercise 6-3. In this exercise, we look at an implementation of the StarGAN trained on a subset of attributes and data.

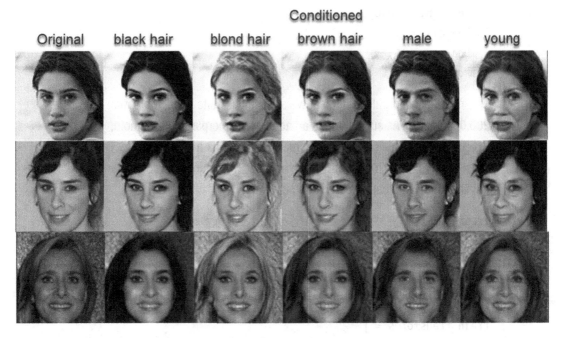

Figure 6-5. *Output from StarGAN trained at 65 epochs*

EXERCISE 6-3. FACE CONDITIONING WITH THE STARGAN

1. Open the GEN_6_StarGAN.ipynb notebook from the GitHub project site. If you are unsure how, then consult Appendix B. Go ahead and run the entire notebook by selecting Run ➤ Run all from the menu.

2. The bulk of this notebook's code should be familiar, but we will again cover the new or different hyperparameters shown. The n_residual_blocks hyperparameter allows for us to control the number of residual blocks used in the generator, where lambda_cls and lambda_rec control the scale on the class and reconversion loss, respectively. The lambda_gp hyperparameter scales the amount of gradient penalty that affects the adversarial loss. The number of critics (n_critics) is the number of times the discriminator trains before the generator.

    n_critic=5,

```
residual_blocks=16,
lambda_cls = 1,
lambda_rec = 10,
lambda_gp = 10
```

3. The image-aligned CelebA dataset we are using is quite large, more than 200,000 images. As such, there is an additional hyperparameter of note, `train_split`, which controls the amount of data the model trains on. The value is set to .2, or 20 percent, of the images. This is done to reduce the training time, but it can affect the output quality.

    ```
    train_split=.2
    ```

4. Another important modification we make is to zoom in on the face images and then crop them when we load them into the dataset. This is performed with the set of transforms shown here and greatly improves training performance:

    ```
    train_transforms = [
            transforms.Resize(int(1.25 * hp.img_size), Image.BICUBIC),
            transforms.CenterCrop(hp.img_size),
            transforms.RandomHorizontalFlip(),
            transforms.ToTensor(),
            transforms.Normalize((0.5, 0.5, 0.5), (0.5, 0.5, 0.5)),
    ]
    ```

5. Next, we will jump down to the loss function definition and the creation of the model's code. At the top of the code the loss function `criterion_cls` takes as input the `logit` (class) and target, which it then passes through the binary cross entropy loss with the logits function. The output of this is divided by `size(0)`, or the batch size, thus averaging the results.

    ```
    def criterion_cls(logit, target):
        return F.binary_cross_entropy_with_logits(logit, target, size_
        average=False) / logit.size(0)

    # Initialize generator and discriminator
    generator = GeneratorResNet(img_shape=input_shape, res_blocks=hp.
    residual_blocks, c_dim=c_dim)
    discriminator = Discriminator(img_shape=input_shape, c_dim=c_dim)

    if cuda:
    ```

```
generator = generator.cuda()
discriminator = discriminator.cuda()
criterion_cycle.cuda()
```

```
generator.apply(weights_init_normal)
discriminator.apply(weights_init_normal)
```

6. Jump down to the training code, and we will first focus on the discriminator or adversarial loss, shown here. The key difference is the addition of the classification loss using the `criterion_cls` loss function.

```
# Real images
real_validity, pred_cls = discriminator(imgs)
# Fake images
fake_validity, _ = discriminator(fake_imgs.detach())
# Gradient penalty
gradient_penalty = compute_gradient_penalty(discriminator, imgs.data,
fake_imgs.data)
# Adversarial loss
loss_D_adv = -torch.mean(real_validity) + torch.mean(fake_validity) +
hp.lambda_gp * gradient_penalty
# Classification loss
loss_D_cls = criterion_cls(pred_cls, labels)
# Total loss
loss_D = loss_D_adv + hp.lambda_cls * loss_D_cls
```

7. Scroll down a little to the generator loss section, and notice how the `n_critic` value sets how frequently the generator is trained in the `if` statement, thus allowing how frequently we apply loss to the discriminator versus the generator. This is a useful hyperparameter to tweak if you feel one or other of the models are surpassing the other.

8. You should also be able to see the generator loss section, as shown here. Again, the loss calculations are like previous exercises, where we first calculate adversarial loss, then classification, and finally cycle loss. Adding them all up using the lambda scale factor.

```
# Translate and reconstruct image
gen_imgs = generator(imgs, sampled_c)
recov_imgs = generator(gen_imgs, labels)
```

```
# Discriminator evaluates translated image
fake_validity, pred_cls = discriminator(gen_imgs)
# Adversarial loss
loss_G_adv = -torch.mean(fake_validity)
# Classification loss
loss_G_cls = criterion_cls(pred_cls, sampled_c)
# Reconstruction loss
loss_G_rec = criterion_cycle(recov_imgs, imgs)
# Total loss
loss_G = loss_G_adv + hp.lambda_cls * loss_G_cls + hp.lambda_rec *
loss_G_rec
```

9. This exercise can take a significant amount of time to train. However, the time is well spent as this example can produce quite realistic results. Be sure to open the `list_attr_celeba.txt` file located in the `images` folder. From this file, you can see a list of attributes you can train this example on; there are about 30 of them. Swap the attribute names out by adjusting the hyperparameter list `selected_attrs`.

As you can see in Figure 6-5, this GAN can produce some impressive and fun results. It may take a while, but training this GAN can be more than worth it. This is an example that you likely could extend and grow on for your own applications. It really is that good.

Before we look at our final GAN of the chapter, we want to take a quick diversion into using existing and established reference models in the next section.

Using the Best with Transfer Learning

Likely by now you started to notice a pattern in our models and the way we work with them. This is, of course, intentional and allows for code to easily be swapped from one exercise to another. In deep learning, we can share code, but we can also share models, the implementation of the code, and/or the trained weights.

When we build a model in PyTorch or another framework, that model is compiled into a mathematical function graph. This graph can then be saved to a file on disk for later reuse without the original code. Furthermore, we can also save the trained parameters/weights of a model for later reuse.

The flexibility of saving the models and weights not only will help us in the future but also allows for various reference networks to be shared. A reference network or model is often one that through literature publication or competition was established as the cutting edge for its time on a particular task.

One such task many reference networks have been developed for was ImageNet. ImageNet is a massive dataset of 2 million plus images defined by 1,000 classes. Networks are trained and tested on ImageNet in terms of classification accuracy, where the most accurate models at the time typically get introduced as a reference model.

It may at first seem odd that we could use models trained on diverse image sets like ImageNet and still be able to reuse such models, until you consider that a deep learning model is constructed in layers, and those layers may be removed or added to and then recompiled. Removing and adding layers to a model allow for it to be extended to new shared domains.

Figure 6-6 shows how we would remove the bottom layers of a model, which are the classification layers, and replace them with new ones, where we also freeze several existing layers, namely, the top layers of a model. Freezing these layers means that we are fixing the weights and not allowing the model to train those any further. The bottom and new layers are set to trainable and allow the model to learn new classification tasks.

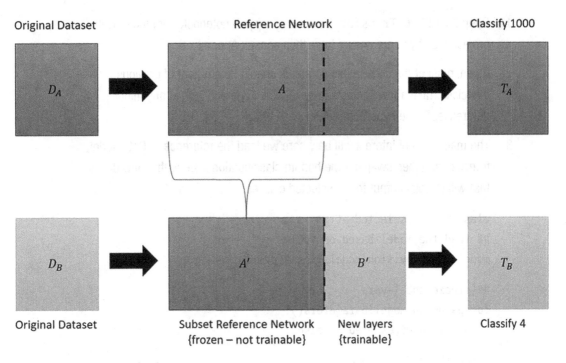

Figure 6-6. *Transfer learning*

We can thus take a reference model like the VGG19 shown in Figure 6-2 and repurpose it from outputting 1,000 classes to one a model that outputs 4. The benefit here is that we reuse the model's trained feature extraction abilities in a new model. Since we are training only a couple new layers in the new model, we can quickly train a new model to classify new output.

Reusing learned feature extraction from one model to another goes only so far. Keep in mind that a model trained to classify cats and dogs may not transfer well to one that needs to identify heavy machinery. Fortunately, reference networks trained on the 1,000-class ImageNet dataset have quite diverse feature extraction that can work across many tasks.

In Exercise 6-4, we are going to look at repurposing a reference VGG19 model to classify against the CelebA dataset. We will use the annotations from the previous exercise with the StarGAN to use as class output from the celebrity faces. Since we are reusing a pretrained reference model, our expectations will be that the model can learn quickly.

EXERCISE 6-4. FEATURE REENGINEERING WITH TRANSFER LEARNING

1. Open the GEN_6_Transfer_Learning.ipynb notebook from the GitHub project site. If you are unsure how, then consult Appendix B.

2. Again, most of the code in this notebook should be a repeat of previous exercises. Take a few seconds to review the code and look at any minor differences in the hyperparameters.

3. The main point of interest will be where we load the reference VGG19 model, freeze it, and then swap out the bottom classification layer with a new design that will classify output to our selected classes.

```
classes = hp.selected_attrs
## Load the model based on VGG19
model = torchvision.models.vgg19(pretrained=True)

## freeze the layers
for param in model.parameters():
    param.requires_grad = False

# Modify the last layer
```

```
number_features = model.classifier[6].in_features
features = list(model.classifier.children())[:-1] # Remove last layer
features.extend([torch.nn.Linear(number_features, len(classes))])
model.classifier = torch.nn.Sequential(*features)
```

4.   And yes, it is that easy. We now have a new model that is trained to extract features that we can repurpose to a new classifier. It is one that works with our selected classes.

5.   Slide down to the training code near the bottom, and notice how we train the model. Notice how we use torch.max to reduce the labels from an array of classes to the maximum. This means we are only training the model to a maximum output of a single class.

```
inputs , labels = data
inputs = inputs.cuda()
labels = labels.cuda()

optimizer.zero_grad()

with torch.set_grad_enabled(True):
    outputs  = model(inputs)
    loss = loss_fn(outputs, torch.max(labels, 1)[1])

loss.backward()
optimizer.step()
```

6.   After the model has completed training, we want to move on and test it against the validation dataset visually. We do that with the following code:

```
num_images = 6
was_training = model.training
model.eval()

with torch.no_grad():
    (inputs, labels) = next(iter(val_dataloader))
    inputs = inputs.cuda()
    outputs = model(inputs)
    preds = torch.where(outputs > .5, 1, 0)
    for i in range(len(inputs)):
      pred_label = [hp.selected_attrs[i] for i,label in
      enumerate(preds[i]) if label > 0]
```

```
truth_label = [hp.selected_attrs[i] for i,label in
enumerate(labels[i]) if label > 0]
print(f"Predicted: {(pred_label[0] if pred_label else None)}
Truth: {truth_label}")
imshow(make_grid(inputs[i].cpu()), size=pic_size)
model.train(mode=was_training)
```

7. Figure 6-7 shows a visual sample of the output from the model. Notice how the model predicts only a single class; this is on purpose since the model was trained to only a single class at a time.

Predicted: Blond_Hair Truth: ['Blond_Hair', 'Brown_Hair', 'Young']

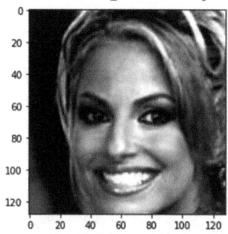

Predicted: Black_Hair Truth: ['Black_Hair', 'Young']

Figure 6-7. *Example predicted output*

8. This model quickly trains, so go back to try different classes and see what effect this has on the predicted output.

Now, you don't always need to freeze all the reference model layers. In fact, you can choose to freeze only a portion of the model allowing some layers to adjust the learned feature extraction to your domain. We typically will unfreeze or not freeze lower layers in the model. These are the layers that are more fine-tuned to specific domains.

Transfer learning is an excellent way to relearn models for a variety of applications from classification to just feature extraction. In the next section, we will explore a new GAN that uses the VGG19 for simply that, feature extraction.

Increasing Resolution with SRGAN

In the 1982 movie *Blade Runner* by director Ridley Scott, there is a scene where the protagonist played by Harrison Ford uses software to zoom in and enhance a static image. While the movie was set in the future, it wasn't long before this same technique was used time and again in movies and TV shows.

It became such a stereotype that comedians and even memes were constructed making fun of this supposed application and explaining that image analysis software could never work that way. Well, that was all before generative modeling and the advent of deep learning.

Fortunately, with the growth of generative modeling and deep learning, we can now do exactly what the futuristic movie made almost 40 years proposed. With generative modeling, we can now zoom in on images and at the same time increase the image resolution.

The GAN that can do this is called the *super resolution* (SRGAN) and was initially designed to just increase the resolution of images. With some slight modifications, we will look at not only how this GAN can increase resolution but also zoom in at the same time.

An SRGAN is somewhat simple but combines the best of a few techniques to do its magic. In the SRGAN, the generator is paired with a feature extractor that determines feature loss. This feature extractor model is based upon a VGG19 model, but instead of repurposing the model, we just remove the last layer.

By removing the last layer of the VGG19 model, we convert the model output from 1,000 classes to just the extracted features in the last layer. These extracted features can then be used to compare against real images to determine further loss or error in generation. This requires the addition of new classification layers that can then learn to reclassify the extracted features.

Combined with the ResNet, the SRGAN becomes a relatively simple but powerful model as we will see in the next exercise. This model starts out training very slowly, so be patient and don't panic if you see poor output for some time. Let's jump into Exercise 6-5 and see what the future holds.

EXERCISE 6-5. INCREASING RESOLUTION WITH THE SRGAN

1. Open the GEN_6_SRGAN.ipynb notebook from the GitHub project site. If you are unsure how, then consult Appendix B.

2. The bulk of this code will be familiar, but look at how we define a new hyperparameter in the section. SRGAN uses a single image set like CelebA but will prepare and resize a copy of the images to our desired output. To set how much we enhance the image, we use hr_size to define the high-resolution image set's size.

```
img_size=128,
hr_size=256,
```

3. Scroll down to the ImageSet class data loader that loads the images. Inside, note how we are creating two sets of images using the first set. One set of images will be low resolution and the other high resolution. The code to transform the various image sets is shown here:

```
self.lr_transform = transforms.Compose(
    [
        transforms.Resize((hr_height // 4, hr_height // 4),
        Image.BICUBIC),
        transforms.ToTensor(),
        transforms.Normalize(mean, std),
    ]
)
self.hr_transform = transforms.Compose(
```

```
    [
        transforms.Resize(int(1.5*hr_height), Image.BICUBIC),
        transforms.CenterCrop(hr_height),
        transforms.ToTensor(),
        transforms.Normalize(mean, std),
    ]
)
```

4. Notice how the transform for the high-resolution images, `self.hr_transform`, includes the zoom-in we used for the StarGAN and a center crop, effectively zooming into the face of the celebrity. `self.lr_transform` downsizes the image by `hr_height // 4`, thus reducing the image to one-fourth of the size.

5. Scroll again down to the model definition and look at the new model class `FeatureExtractor` shown here. Inside this model we can see how the VGG19 reference model is created with a pretrained model. Then the model is sliced from the 18th layer, thus removing that last classification layer.

```
class FeatureExtractor(nn.Module):
    def __init__(self):
        super(FeatureExtractor, self).__init__()
        vgg19_model = vgg19(pretrained=True)
        self.feature_extractor = nn.Sequential(*list(vgg19_model.
        features.children())[:18])

    def forward(self, img):
        return self.feature_extractor(img)
```

6. Next, we can scroll down a little and take a quick look at how the three models, generator, discriminator, and feature extractor, are created.

```
generator = GeneratorResNet()
discriminator = Discriminator(input_shape=(hp.channels, *hr_shape))
feature_extractor = FeatureExtractor()

# Set feature extractor to inference mode
feature_extractor.eval()
```

7. Zoom down to the training code block and see how the loss is calculated for the generator. Notice how the content loss is used in the determination of final loss.

```
# Generate a high resolution image from low resolution input
gen_hr = generator(imgs_lr)

# Adversarial loss
loss_GAN = criterion_GAN(discriminator(gen_hr), valid)

# Content loss
gen_features = feature_extractor(gen_hr)
real_features = feature_extractor(imgs_hr)
loss_content = criterion_content(gen_features, real_features.detach())

# Total loss
loss_G = loss_content + 1e-3 * loss_GAN
```

8. This model can take a while to train, and the initial output can appear like
    garbage. Be patient and wait for the model to get a few epochs in before the
    magic starts to happen.

Figure 6-8 shows the results of the model after training for a few hours. The leftmost
images show the original 64×64 low-resolution images, and the right side shows the high-
resolution 256×256 images. While it looks like all that is done is zooming in, consider that
the left images are low resolution and being zoomed in on and upscaled at the same time.

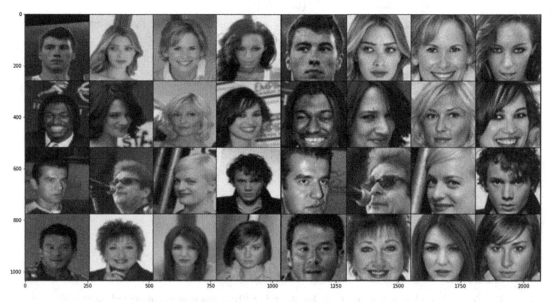

Figure 6-8. *Sample training output from SRGAN*

Generative modeling and the SRGAN have allowed us to do what were only thought possible in science fiction but is now a reality. This will surely be shaped by the explosion of generative modeling and GANs like the SRGAN. Whatever the outcome, the possibilities with the SRGAN remain to be seen.

Conclusion

ResNets applications in GANs have allowed the generative models to now surpass some of our wildest expectations. The GANs we looked at in this chapter can certainly be used out of the box to produce some realistic content from performing gender swaps to changing hair color.

At this stage of the book the models we develop could be used for a wide range of commercial applications from doing simple image enhancements to showing a customer what a different hair color or outfit may look like. The possibilities are endless and powerful and only getting more so every day.

As we progress through this book, we will continue to focus on realism, but this chapter should be a milestone. It likely may be the first time you were able to generate a very real-looking fake image. It can be both exciting and upon further reflection frightening to think of the possibilities.

With these newfound skills you find in this book, it is also important to remember that you are becoming a wizard of a new age. While this collective group of AI wizards does not have established code, there are several yet unwritten rules.

Of these unwritten or partially written rules, one rises to the top in almost every list, which is to do no harm with the AI models we build. For many, this is the golden rule, and many companies from Google to Microsoft have seen widespread staff dissention if they attempt to break this rule.

Therefore, I hope that the knowledge you gain in this and successive chapters is used only for good. Always keep this in mind when building your models. While you can't think of every application, consider the possible ways someone could do harm and avoid them.

In the next chapter, we look at further ways of enhancing our generative models with the advent of a new mechanism called *attention*. Attention is currently taking the deep learning world by storm and injecting itself into applications from natural language processing to of course GANs.

## CHAPTER 7

Attention Is All We Need!

The word *attention* is derived from the Latin *attentionem*, meaning to give heed to or require one's focus. It's a word used to demand people's focus, from military instructors to teachers and parents. Yet, it is a word we also use in mathematics and computer science to describe how well something attends or links to something else.

In 2017, the pivotal paper "Attention Is All You Need" took this concept a step further and introduced the application of attention to deep learning for the purpose of understanding how well features attend to other features. The authors of the paper were able to show instrumental results that have altered deep learning since.

While in the original paper the focus of showing a working attention mechanism was used in the application of natural language processing, it wasn't until a year later when the creator of the GAN demonstrated a working attention mechanism within a GAN called the *self-attention GAN* (SAGAN).

In this chapter, we will take a close look at how an attention mechanism works and how it can be incorporated into deep learning. We first look at the various types of attention and how they are used. From there we look at how the basic attention mechanism works to extract feature relationships using a code example. We finish by looking at an attention visualization example that works with convolutional layers.

Introducing the self-attention mechanism adds training stresses, and before we jump to using attention, we need to understand some fundamental properties of functions. One such property is Lipschitz continuity, which is fundamental to balanced training in GANs. We therefore spend some time understanding what Lipschitz is and how it can be used.

Next, we come to building an example of an SAGAN that is trained on the CelebA faces dataset where we train a model to blindly generate real faces from nothing. Finally, we finish the chapter by building an SAGAN model with residual networks and several Lipschitz constraint helpers.

© Micheal Lanham 2021
M. Lanham, *Generating a New Reality*, https://doi.org/10.1007/978-1-4842-7092-9_7

You have worked hard to get here, and in this chapter, we start to see huge rewards pay off in the models we build and use. The models we build or use now produce cutting-edge results, which should make this our most exciting chapter yet. We will cover the following in this chapter:

- Attention

- Augmenting convolution with attention

- Lipschitz continuity in GANs

- Building the self-attention GAN

- Improving on the SAGAN

This chapter will likely feel like a whirlwind compared to the previous few chapters you worked through to get here. We will spend more time understanding at the core how to effectively train GANs using fundamental mathematical properties. We also dig into how context and relationships are identified in feature mapping.

What Is Attention?

We can think of the word *attention* as meaning how well you or another entity focus on a specific task or another entity. In the military, you may be asked to come to attention, meaning to stand straight and focus forward, while a close friend or relative may use the word as a plea for your companionship and focus.

Either way, we can think of attention as meaning how well one entity focuses on or takes direction from another entity. In simple terms, we can relate objects by saying A attends to B, but B may not attend to A. In other words, A takes direction from B, but B doesn't take direction from A.

Figure 7-1 shows a heat attention map of mouse movement over a web form. In the figure, the colors represent heat or the amount of attention. Red areas represent the hot areas of focus for our attention, or where visitors hovered the mouse over those areas. Colder blue areas represent little or no focus by the mouse.

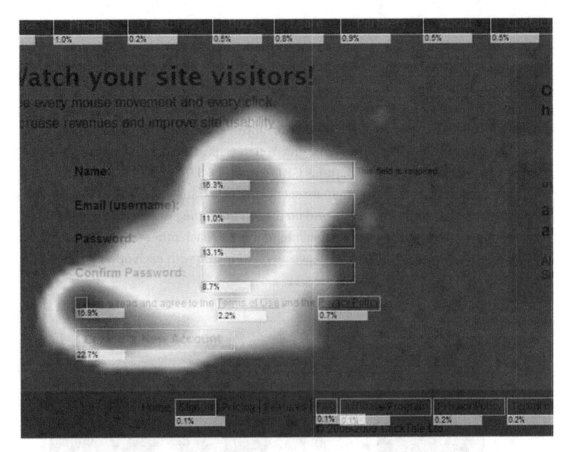

Figure 7-1. *Mouse movement attention map*

Attention used in various applications of mathematics and deep learning often looks very much like the example in Figure 7-1. In fact, we will view some of these attention maps later in this chapter. These examples may use different gradients to represent attention, but the concept is the same.

The idea of attention was first applied to natural language processing as a means of relating word pairs in sequence to sequence translation models. Seq2Seq models, as they are known, are remarkably similar in concept to autoencoders. They effectively translate language by learning from word pairings, an idea not unlike our image to image translation GANs.

Traditionally in NLP applications, recurrent or gated network layers were used to identify relationships between words. This unlike convolution where recurrent networks relies on a recursive method of unfolding sequences to train network layers. This allowed for deep learning models to learn sequences in time or language.

197

Unfortunately, recurrent networks don't scale very well, especially as the learned sequences increase in size. In fact, one of the primary improvement's attention promised in the original paper was an increase in training scalability, especially when applied to longer sequences.

The addition of attention to NLP models is typically wrapped in an upscaled Seq2Seq model type called *transformers*. As a result, transformers with attention radicalized natural language models to unforeseen gains. You may have already heard these models referred to as BERT or the more infamous GPT-2 and GPT-3.

Figure 7-2 shows an attention map for set of language translation text. At the left side of the figure are the words in English, and on the top the same phrase is in Spanish. The brighter areas of the plot show words that attend more to each other than other words. Dark areas indicate where words likely have no relationship with each other.

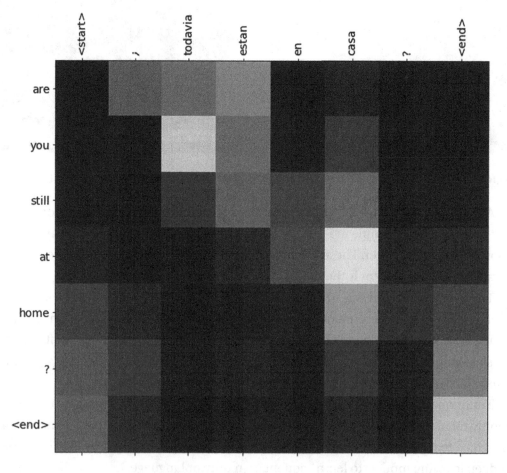

Figure 7-2. *NLP attention map for language translation*

An NLP model uses the attention to then weigh the most likely pairing of words when generating text translations. This allows the model to not only learn word sequences but how words link or relate to other words. This concept has been an effective replacement for the reliance of recurrent networks to NLP tasks.

The GPT-2 took almost a year to release to the public over fears this model would allow for believable yet fake text in the form of news generation. While this book doesn't cover NLP text generation, just realize that these models share many of the same concepts. In the next section, we look at the types of attention and how they relate to each other.

Understanding the Types of Attention

We typically think of the word *attention* as binary, meaning you are focusing or not, though we understand and use various types of focus in our daily lives. We may use other phrasing to suggest our focus is concentrated on a specific problem, local or global, even going so far as focusing on specific elements in sets of tasks or items and understanding their relationship with each other and themselves.

In machine learning, attention is a mechanism we can use to relate features to other features in various contexts. We may want to consider how localized groups of features attend to each other or how a feature is represented in a global context. Attention can be either local or global attention and can be either hard or soft.

Figure 7-3 shows two types of attention being applied to an image. On the left side we see that the head of the dog is in focus and the remainder of the body is still slightly visible but less in focus. This is an example of global attention since we can see the dog's head in relation to the rest of the image. Yet it is also an example of softened attention because we can visualize the loss of focus across the image.

Conversely, the right side of Figure 7-3 shows an example of hard local attention. It's hard because we focus on one localized area and totally ignore the rest of the image. Likewise, this attention is also local since we have no relation to how the dog's head should appear globally in relation to other parts of the image.

Convolutional layers are an example of a localized soft or hard attention mechanism. CNNs are hard focused when used with pooling layers since they remove spatial relationships, thus making the extracted feature output blocky. ResNet models using convolutional residual blocks are an example of a local soft attention mechanism. The softening occurs because of the residuals being passed or skipped across the layers.

Global soft attention   Local hard attention

Figure 7-3. *Examples of types of attention working on an image*

Self-attention, the type of attention suggested in the original paper, is a further type that uses a global soft mechanism to learn feature importance with itself or features within itself. This produces output in the form of feature/attention maps showing the relation from one feature to the other.

Figure 7-4 shows how self-attention may be applied across words in a sentence. The highlighted (red) text shows the focus word in the sentence and how it relates to other words for each set of words in the sentence. The darker the shading around the word, the more related or attended the word is to the focused word. Figure 7-2 also shows an example of self-attention applied to a set of words on an attention map.

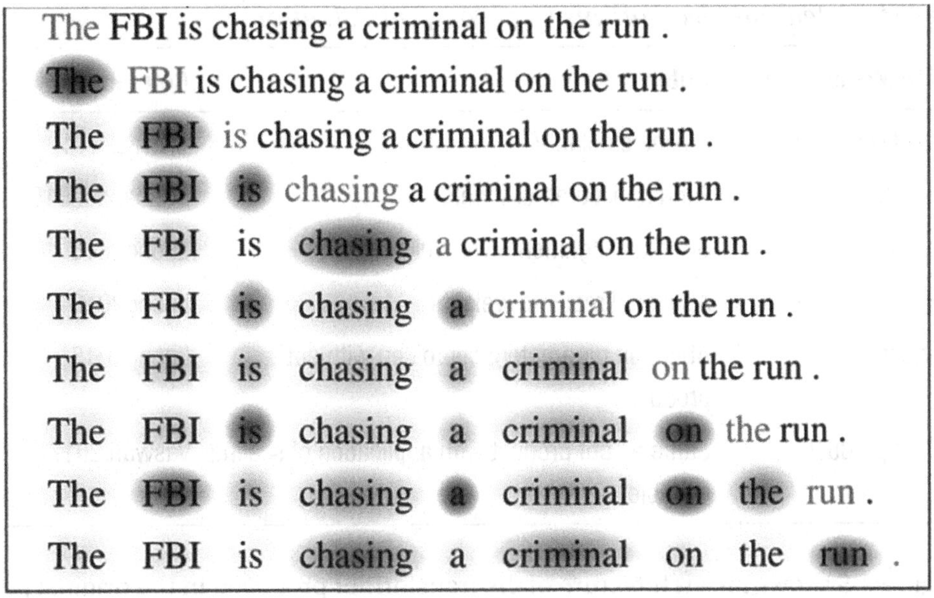

The **FBI** is chasing a criminal on the run .

The FBI is chasing a criminal on the run .

The **FBI** is chasing a criminal on the run .

The **FBI** **is** chasing a criminal on the run .

The FBI is **chasing** a criminal on the run .

The FBI is chasing **a** criminal on the run .

The FBI is chasing a **criminal** on the run .

The FBI **is** chasing a criminal **on** the run .

The **FBI** is chasing **a** criminal **on** **the** run .

The FBI is **chasing** a **criminal** on the **run** .

Figure 7-4. *Self-attention applied to text showing relationship of words in a sentence*

So now that we understand the basic types of attention, we need to look at how we can apply attention in the next section.

Applying Attention

Not only does attention come in various forms from local hard to global soft, it can also be applied with several mechanisms. Table 7-1 shows various examples of attention mechanics used to apply attention for feature extraction and/or relationships. While we could review each method in more detail, our primary focus will be on the last one, transformers.

Table 7-1. *Attention Mechanisms Used to Apply Attention*

Attention Mechanism	Notes	Reference
Content-based	Cosine similarity	Graves2014
Concat or additive	Tanh applied across weights	Bahdanau2015
Location base	Softmax alignment over weights	Luong2015
General	Applied directly to weights	Luong2015
Dot product	Attention parameters combined with dot product	Luong2015
Scaled dot product	Same as dot product with application of scaling variable	Vaswani2017

Figure 7-5 shows the architecture of the transformer proposed in the original paper. In the figure, you can see two forms of attention being used, multihead and scaled dot product. Both combine to provide the self-attention mechanism we discussed earlier.

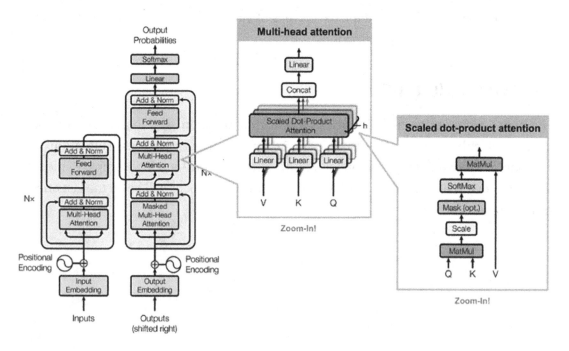

Figure 7-5. *Transformer architecture showing multihead attention*

The multihead attention mechanism was introduced by the authors of the AAYN paper and internally unfolded into the form of scaled dot-product attention shown in the inset of Figure 7-5. Again, remember that the focus of that paper was training NLP models, so the architecture pictured is for NLP transformers and not the GAN design we will see later. However, the multiheaded attention mechanism used to extract self-attention is something we will focus on.

Refer to Figure 7-5 and the far-right zoom-in that highlights the scaled dot-product attention mechanism being applied across three inputs: Q, K, and V. These letters are shortened to Q for querys (yes, the spelling is incorrect), K for keys, and V for values. Q and K are first multiplied and then scaled and passed through a softmax function. The output of this is then multiplied by the input values.

To demonstrate how self-attention is calculated in this manner, we of course need to look at some code in an exercise. Exercise 7-1 doesn't do anything aside from show the steps in the calculation. All inputs and weights are randomized, and you can think of this as the initial phase of training. Let's jump in and see how multihead attention works next.

EXERCISE 7-1. UNDERSTANDING MULTIHEADED ATTENTION

1. Open the GEN_7_Attention.ipynb notebook from the GitHub project site. If you are unsure how, then consult Appendix B.

2. The first cell denotes the imports and an example randomized input, as shown here:

```
import torch
import numpy as np

x = np.random.randint(0,3,(3,4))
x = torch.tensor(x, dtype=torch.float32)
print(x)
```

3. Again, the inputs and weights are randomized, and we only want to focus on the calculation steps. Here is where we create the randomized weights for the query, key, and values:

```
w_key = np.random.randint(0,2,(4,3))
w_query = np.random.randint(0,2,(4,3))
w_value = np.random.randint(0,4,(4,3))
```

```
w_key = torch.tensor(w_key, dtype=torch.float32)
w_query = torch.tensor(w_query, dtype=torch.float32)
w_value = torch.tensor(w_value, dtype=torch.float32)
print(w_key)
print(w_query)
print(w_value)
```

4. In a real multiheaded self-attention layer, the weights would be trained over time.

5. Next, we apply/multiply all the weights to the input x to derive keys, querys, and values.

```
keys = x @ w_key
querys = x @ w_query
values = x @ w_value

print(keys)
print(querys)
print(values)
```

6. After that, we multiply querys by the transpose of the keys and find the attn_scores value.

```
attn_scores = querys @ keys.T
print(attn_scores)
```

7. Then we apply a softmax function to the output attn_scores to get attn_scores_softmax.

```
from torch.nn.functional import softmax

attn_scores_softmax = softmax(attn_scores, dim=-1)
print(attn_scores_softmax)
```

8. We finish off by generating the outputs by multiplying by the values and summing the results.

```
weighted_values = values[:,None] * attn_scores_softmax.T[:,:,None]

outputs = weighted_values.sum(dim=0)
print(outputs)
```

The main thing you should take away from the previous exercise is the functional steps in determining how to apply multiheaded attention to an input x using a simple example. An actual implementation of multiheader attention with convolution is more complicated and something we will discuss in the next section.

Augmenting Convolution with Attention

The concept of attention (self-attention) augmented convolution was first introduced through the self-attention GAN paper. In the paper, the authors showed how convolutional layers could be augmented using a multihead self-attention mechanism.

Figure 7-6 is extracted from the original SAGAN paper and demonstrates how a single augmented attention layer works. From the left input side, the convolution outputs enter the mechanism and are split into 1×1 convolutional layers. The out f(x) represents the keys head, the g(x) function represents querys, and of course h(x) represents values.

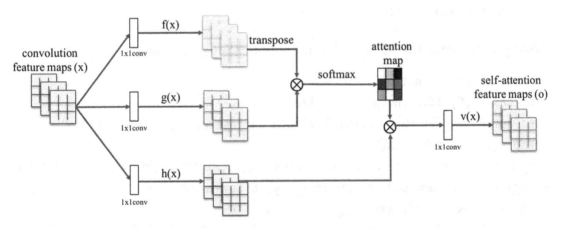

Figure 7-6. *Multiheaded self-attention applied to convolution layer*

Recall from Exercise 7-1 how the rest of the calculations are applied, and you have a high-level understanding of how multiheaded, self-attention is calculated. The output self-attention feature maps are then passed into successive convolutional blocks. How much and where you apply self-attention is up to the model architecture. In most cases, you will apply self-attention to the lower and/or output layers of a model.

Listing 7-1 is extracted from the Self_Attn layer class from the GEN_7_SAGAN. ipynb example. The Self_Attn class provides a wrapper for convolutional augmented self-attention. We will look at the full example later in the chapter, but notice how the query and key convolutional layers alter the input in_dim by dividing by 8, but the value convolution remains the same.

Listing 7-1. Attention Convolutional Layers

```
self.query_conv = nn.Conv2d(in_channels = in_dim , out_channels = in_dim//8 ,
kernel_size= 1)
self.key_conv = nn.Conv2d(in_channels = in_dim , out_channels = in_dim//8 ,
kernel_size= 1)
self.value_conv = nn.Conv2d(in_channels = in_dim , out_channels = in_dim ,
kernel_size= 1)
```

Within the forward function of the Self_Attn class, we can see where the self-attention mechanism is applied. Aside from the tensor manipulations with view and the use of torch.bmm, which stands for batch matrix multiply, the code shown in Listing 7-2 is like what we covered in Exercise 7-1.

Listing 7-2. Inside the Forward Function of the Self_Attn Class

```
#inside forward function
m_batchsize,C,width ,height = x.size()
proj_query  = self.query_conv(x).view(m_batchsize,-1,width*height).
permute(0,2,1) # B X CX(N)
proj_key =  self.key_conv(x).view(m_batchsize,-1,width*height) # B X C x (*W*H)
energy =  torch.bmm(proj_query,proj_key) # transpose check
attention = self.softmax(energy) # BX (N) X (N)
proj_value = self.value_conv(x).view(m_batchsize,-1,width*height) # B X C X N

out = torch.bmm(proj_value,attention.permute(0,2,1) )
out = out.view(m_batchsize,C,width,height)

out = self.gamma*out + x
return out,attention
```

Notice that the forward function in Listing 7-2 has two outputs, the attended output out and the calculated `attention` map. While we won't project those outputs any further, what is possible is visualizing those attention maps. We can work through Exercise 7-2 to visualize how those attention maps look next.

EXERCISE 7-2. VISUALIZING SELF-ATTENTION MAPS

1. Open the `https://epfml.github.io/attention-cnn/` in your `browser` notebook from the GitHub project site.

2. This site was developed as part of the paper "On the Relationship between Self-Attention and Convolutional Layers" published at ICLR 2020.

3. Figure 7-7 shows the start of the visualization table and provides the options for visualizing three types of attention.

Select attention type:

RELATIVE SELF-ATTENTION	POSITION-ONLY SELF-ATTENTION	VISION TRANSFORMER
Use 2D relative positional encoding and image content to compute the attention.	Discard the pixel values and compute the attention scores only on relative positions.	Use absolute 1D positional encoding and CLS token for classification. ViT-Base/16.

Select image and query pixel:

Figure 7-7. *Interface for controlling visualization options*

4. Click the Vision Transformer at the far right, and notice how the images also change.

5. Hover your mouse over the images, and you will see how the attention maps in the table below update as you cross each pixel.

6. Figure 7-8 shows an example of self-attention maps from hovering the pixel over the picture of the shark.

Visualize attention per layer and head

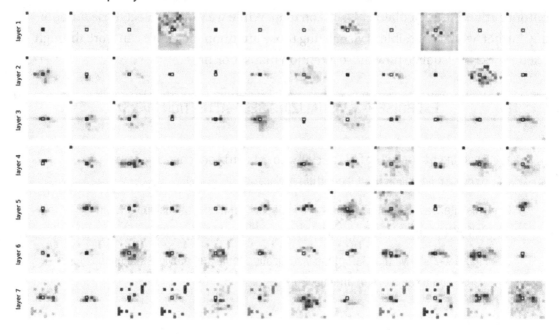

Figure 7-8. *The generated attention maps per layer and head*

7. Look closely at the maps, and in many of them you will see the outline of the shark representing how the pixel attends to the rest of the features on the shark.

The web page and paper demonstrate multiheaded attention over each layer of a model. In some cases, you may want to apply attention across every layer in a model, and in others you may not.

As of 2020, augmented attention for convolution is becoming quite popular and for good reason. An attention mechanism is a powerful tool to provide learned context to the features we extract with convolution or other means. However, we always need to balance the improvements over either side to build a good generator. So, before we apply attention to a GAN, we want to review again the importance of training balance in the next section.

Lipschitz Continuity in GANs

For most of this book we have discussed the need to balance the training of the generator and the discriminator. Without balance, if either side ever becomes too accurate, then the other has no hope of ever getting better.

As an example, imagine you are training to be a champion runner. Graciously your training competition is Usain Bolt, fastest man on Earth. Every day you train and measure your improvements next to Usain. Unfortunately, Usain is also training daily and improves just as much as you if not more. Since you are comparing your performance to Bolt's performance, your gains and therefore your performance begin to diminish. In the end, you quit and just decide to watch from the sidelines.

This same principle works in a regular GAN. Since the generator measures its performance with respect to the discriminator, if the discriminator becomes great at picking real/fake, the loss pushed back to the generator will diminish. In turn, the loss the generator uses to train reduces, causing a vanishing gradient problem.

If you are training a GAN and notice the output starting to degrade, then this is likely a sign that the discriminator is getting too good. The opposite occurs when your generator seems to stop or just not improve anymore. This is a sign the discriminator is regressing or has just reached its maximum potential.

Balance is therefore key to training a GAN, and we have looked at some methods previously to better manage this problem. Recall when we looked at the various methods of measuring and comparing distributions in Chapter 4. We covered the WGAN, a GAN that used an earth mover's distance algorithm to determine loss rather than a divergence function like KL.

Since the balance point in a GAN is the discriminator, we often first look to it to improve balance and training stability. We can do this by reviewing an abstract property of the function the discriminator is trying to approximate to. That property is called the Lipschitz continuity and is what we will cover in the next section.

What Is Lipschitz Continuity?

Lipschitz continuity is a mathematical property of a function that defines a subclass with uniform continuity. That is a function of Lipschitz continuity if it is uniformly continuous across its entire graph. This means the slope for every pair of points on the graph must be within some value or constant.

Figure 7-9 shows two functions, $f(x) = sin(x) \wedge f(x) = x^2$. The first function is Lipschitz continuous since the bounding lines that demonstrate the continuity of change don't intersect the graph. Alternatively, the second function is not Lipschitz continuous since the bounding lines do intersect the graph when the Lipschitz constant is equal to 1.0. While we could increase the constant/slope, our preference for stability is to remain at 1.

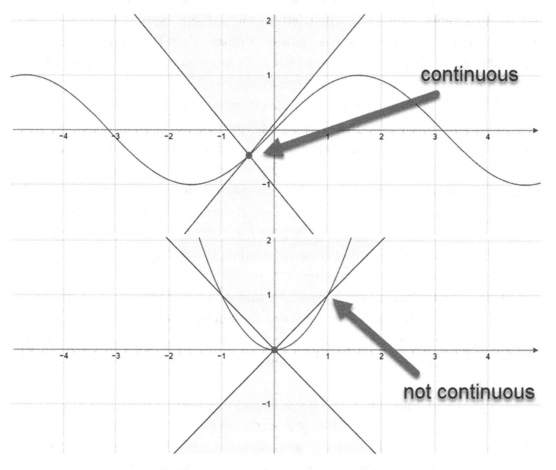

Figure 7-9. *Example of continuous and discontinuous Lipschitz functions*

In the WGAN, with the use of the earth mover's function, the discriminator tries to stay Lipschitz continuous since the function approximates a continuous function. However, remember since our network is approximating a function, it could infer one that still breaks discontinuity.

We can control the LS continuity of a function through a Lipschitz constraint or a measure to control how much a function changes using a few different forms. The constraint is called a *1-Lipschitz constraint* since we look to enforce the change to a constant of 1. The forms listed here are what we typically consider when employing some form of constraint training gradients:

- **"Soft" 1-Lipschitz constraint**: The gradient function is forced to equal or be close to 1 on average. Thus, some function points may have high values, while others low.

- **"Hard" 1-Lipschitz constraint**: This forces the gradient to be lower or equal to one at every point. This typically works well for adversarial training.

- **"Gradient" 1-Lipschitz constraint**: This forces the gradient to be almost 1 everywhere. This is not typically used in other ML areas but is required when training over Wasserstein distance.

To enforce 1-Lipschitz constraints over gradients, we typically use two methods, weight clipping and spectral normalization. Weight clipping, or penalizing the gradient, ensures that weights are clipped such that the Lipschitz constant is less than 1. We have used the gradient_penalty function shown in listing 7-3 throughout the last few chapters for weight clipping.

Listing 7-3. gradient_penalty Function

```
#@title HELPER FUNCTION - COMPUTE GRADIENT PENALTY
def compute_gradient_penalty(D, real_samples, fake_samples, labels):
    """Calculates the gradient penalty loss for WGAN GP"""
    # Random weight term for interpolation between real and fake samples
    alpha = FloatTensor(np.random.random((real_samples.size(0), 1, 1, 1)))
    # Get random interpolation between real and fake samples
    interpolates = (alpha * real_samples + ((1 - alpha) * fake_samples)).
    requires_grad_(True)
    d_interpolates = D(interpolates, labels)
    fake = Variable(FloatTensor(np.ones(d_interpolates.shape)), requires_
    grad=False)
    # Get gradient w.r.t. interpolates
    gradients = autograd.grad(
```

```
        outputs=d_interpolates,
        inputs=interpolates,
        grad_outputs=fake,
        create_graph=True,
        retain_graph=True,
        only_inputs=True,
    )[0]
    gradients = gradients.view(gradients.size(0), -1)
    gradient_penalty = ((gradients.norm(2, dim=1) - 1) ** 2).mean()
    return gradient_penalty
```

The problem with weight clipping in this manner is that approximated functions always assume a Lipschitz constant less than 1. This means that the approximated function is more gradual with smoother hills and valleys. This in turn can wash out important details in the replicated output we are looking to produce.

Spectral normalization is a method that relies on a singular value decomposition of the gradient tensor that produces a sigma value for each element. By taking the largest sigma value of the tensor and dividing it by all values, we can be assured that the maximum is 1, confirming that the function is 1-Lipschitz continuous.

Listing 7-4 shows the SpectralNorm class's compute_weight function we will use as a method to constrain gradients when training. This function uses SVD to determine the value of sigma and then divides all the weights by sigma to assure none is greater than 1.

Listing 7-4. Spectral Normalization compute_weight Function

```
def compute_weight(self, module):
    weight = getattr(module, self.name + '_orig')
    u = getattr(module, self.name + '_u')
    size = weight.size()
    weight_mat = weight.contiguous().view(size[0], -1)
    with torch.no_grad():
        v = weight_mat.t() @ u
        v = v / v.norm()
        u = weight_mat @ v
        u = u / u.norm()
    sigma = u @ weight_mat @ v
    weight_sn = weight / sigma
```

We can see how both methods work to help improve discriminator training and the balance of a GAN in the next exercise. While both methods can work separately as we have seen with gradient penalties in previous chapters, but Exercise 7-3 demonstrates both.

EXERCISE 7-3. LIPSCHITZ CONTINUITY IN GANS

1. Open the `GEN_7_Lipschitz_GAN.ipynb` notebook from the GitHub project site. Run the entire notebook.

2. This code example is nearly identical to the Chapter 4 `GEN_4_DCGAN.ipynb` exercise notebook. The only difference here is that gradient penalty loss and spectral normalization have been added.

3. Adding gradient penalty loss and spectral normalization is relatively easy. The following code shows a discriminator and internal convolution block with the `SpectralNorm` class wrapping the convolutional layer.

```
class Discriminator(nn.Module):
  def __init__(self):
    super(Discriminator, self).__init__()

    def discriminator_block(in_filters, out_filters, bn=True):
      block = [SpectralNorm(nn.Conv2d(in_filters, out_filters,
      3, 2, 1)),
                nn.LeakyReLU(0.2, inplace=True), nn.Dropout2d(0.25)]
      if bn:
          block.append(nn.BatchNorm2d(out_filters, 0.8))
      return block
```

4. We could just as easily add spectral normalization to the generator convolutional layers using the `SpectralNorm` class. The generator is kept clean here to focus on the discriminator.

5. If we move down to the training block, we can again review how the gradient penalty loss is applied to the discriminator loss with the following:

```
real_loss = loss_fn(discriminator(real_imgs), valid)
fake_loss = loss_fn(discriminator(gen_imgs.detach()), fake)
```

```
# Gradient penalty
gradient_penalty = compute_gradient_penalty(discriminator, real_imgs.
data, gen_imgs.data)

loss_D = real_loss + fake_loss + hp.lambda_gp * gradient_penalty
```

6. `lambda_gp` is the scale hyperparameter that adjusts how much of the gradient penalty to apply to the total calculation of loss. You can refer to the `gradient_penalty` function by scrolling up a few blocks in the notebook.

7. This notebook will generally take much longer to stabilize, so the output may appear quite random at start. Just realize that is a consequence of maintaining Lipschitz continuity. Be patient and let the sample train for a few hours to notice the increase in the number of details generated.

Now that we have completed our divergence on Lipschitz continuity in GANs, we return to our work with self-attention and the self-attention GAN in the next section.

Building the Self-Attention GAN

What the authors of the original SAGAN paper discovered was that the self-attention mechanism caused the discriminator to lose Lipschitz continuity even when trying to balance training with a gradient penalty and/or Wasserstein distance. They therefore implemented a fix using spectral normalization in a form we just covered in the previous section.

By incorporating self-attention and the required constraints, aka spectral normalization, the authors were able to convert a simple DCGAN into a SAGAN. This in turn produced some incredible results for straight blind sample generation, not image-to-image translation.

In Exercise 7-4, we look at a base implementation of the SAGAN trained to generate faces from the CelebA dataset. Keep in mind that this generator (GAN) is completely blind generating faces just using convolution with self-attention. Of course, to balance the training, we throw in some spectral normalization to avoid things getting crazy.

EXERCISE 7-4. BUILDING AND TRAINING THE SAGAN

1. Open the GEN_7_DCGAN_SAGAN.ipynb notebook from the GitHub project site.
    Run the entire notebook.

2. This code example is nearly identical to the Chapter 4 GEN_4_DCGAN.ipynb
    exercise notebook. The key differences here are the addition of self-attention
    and spectral normalization in the models.

3. We can see how the self-attention layers are added to the models by scrolling
    down to the Generator class definition and viewing the layer configuration
    and the start of the forward function shown here:

```
self.attn1 = Self_Attn( 512, 'relu')
self.attn2 = Self_Attn( 256, 'relu')
self.attn3 = Self_Attn( 128, 'relu')
self.attn4 = Self_Attn( 64,  'relu')

def forward(self, z):
  z = z.view(z.size(0), z.size(1), 1, 1)
  out=self.l1(z)
  out,_ = self.attn1(out)
  out=self.l2(out)
  out,_ = self.attn2(out)
  out=self.l3(out)
  out,p1 = self.attn3(out)
  out=self.l4(out)
  out,p2 = self.attn4(out)
  out=self.last(out)

  return out, p1, p2
```

4. Likewise, self-attention can be added to the Discriminator class in the
    same manner as shown here:

```
self.attn1 = Self_Attn(64, 'relu')
self.attn2 = Self_Attn(128, 'relu')
self.attn3 = Self_Attn(256, 'relu')
self.attn4 = Self_Attn(512, 'relu')
```

```
def forward(self, x):
  out = self.l1(x)
  out,p0 = self.attn1(out)
  out = self.l2(out)
  out,p0 = self.attn2(out)
  out = self.l3(out)
  out,p1 = self.attn3(out)
  out=self.l4(out)
  out,p2 = self.attn4(out)
  out=self.last(out)
  return out.squeeze(), p1, p2
```

5. All the layers in this model use self-attention in between. However, you can adjust this behavior by simply commenting out the attention layers you don't want in the model. Adding attention layers decreases training performance due to extra calculations and the increased need to balance Lipschitz continuity.

6. If you want, you can comment out some of all the attention layers to see what effect this has on training. If you do so, be sure to adjust the attention outputs of the model (p1, p2).

7. Finally, in the training block, we can see how this is all pulled together to calculate loss, as shown here:

```
d_out_real,dr1,dr2 = discriminator(real_images)
#hinge loss
d_loss_real = torch.nn.ReLU()(1.0 - d_out_real).mean()

z = tensor2var(torch.randn(real_images.size(0), hp.latent_dim))
fake_images,gf1,gf2 = generator(z)
d_out_fake,df1,df2 = discriminator(fake_images)

#hinge loss
d_loss_fake = torch.nn.ReLU()(1.0 + d_out_fake).mean()

# Backward + Optimize
d_loss = (d_loss_real + d_loss_fake) / 2

d_loss.backward()
optimizer_D.step()
```

8. In this example, we use hinge loss to determine the loss by passing the output through a ReLU function and taking the mean. This is just a means for controlling negative loss. Notice how the attention map outputs from the discriminator are not used.

9. Figure 7-10 shows the output of training this exercise for a few hours (Epoch14). Notice the quality of the output, and keep in mind that all the faces are not real people. These faces are generated completely blind.

Figure 7-10. *Example output from training exercise GEN_7_DCGAN_SAGAN.*
ipynb

The output of the previous exercise is quite impressive when you consider the generator is completely working blind when generating the faces. This contrasts with the image to image models we looked at in the past couple chapters.

From here we can only go up, as they say, and in the next chapter we will look to improve on this example to look at generating conditional faces while also incorporating a residual network for better sample generation.

Improving on the SAGAN

Self-attention provides another aspect of feature learning, that of globalized feature localization over convolution, but as we learned this doesn't come without side effects. These side effects can be minimized by enforcing Lipschitz continuity, but we still have other problems like feature over feature extraction.

To help mitigate these issues, we are going to look at another example of an SAGAN that uses residual blocks for skip connections in convolutions as well as adds conditional generation. As we saw earlier, conditional generation and discriminating can improve performance by localizing subdomains.

For Exercise 7-5, we look at improving on the SAGAN using a ResNet convolutional model, labels, and conditions. We will again use the CelebA dataset, this time using the labeled attributes as classes for each picture. As such, we want our classes to be unique and not shareable across images. To do that, we will stick to using the basic hair colors (blond, black, and brown) as classes and only load celebrity faces that are attributed as such. Exercise 7-5 improves on our previous work in that it also attempts to generate 128×128-pixel faces, the largest blind generating we have done thus far.

EXERCISE 7-5. IMPROVING THE SAGAN

---

1. Open the GEN_7_Celeb_SAGAN.ipynb notebook from the GitHub project site. Run the entire notebook.

2. This code example is again based on the Chapter 4 GEN_4_DCGAN.ipynb exercise notebook. Most of the new code resides in just a couple blocks where the models and helper functions are defined. Note that this notebook currently uses all the memory of Google Colab when using a GPU runtime. As such, you need to be careful about how you modify some of the hyperparameters like batch size or image size.

    Increasing the batch size of any model increases the memory requirements of the model as it is processing the forward pass. By reducing the batch size, you reduce the memory requirements, but this can degrade training performance quality and increase running time.

3. Jump down to the model's section and the Generator class definition. Notice how the SelfAttention class is used in between a couple of the layers:

```
ConvBlock(512, 512, n_class=n_class),
ConvBlock(512, 256, n_class=n_class),
SelfAttention(256),
ConvBlock(256, 128, n_class=n_class),
SelfAttention(128),
ConvBlock(128, 64, n_class=n_class)])
```

4. Likewise, we can see the same structure in the discriminator construction:

```
SelfAttention(128),
conv(128, 256, downsample=False),
SelfAttention(256),
conv(256, 512),
conv(512, 512),
conv(512, 512))
```

5. Notice the doubling up of the 512 layers in both the generator and discriminator model construction. You could play with adding more duplicate layers or self-attention layers. However, this current configuration uses the maximum GPU memory. While you could swap the runtime to CPU and increase the model parameters, training such a model on a CPU would be considerably more time-consuming.

6. One other new feature of this example is the introduction of a scheduler. Schedulers allow us to modify hyperparameters at runtime (that is, those that can be modified). While you can schedule any hyperparameter to change, during training the usual suspect is the learning rate. In the optimizers section, we create two new schedulers like so:

```
scheduler_G = StepLR(optimizer_G, step_size=1, gamma=hp.lr_gamma)
scheduler_D = StepLR(optimizer_D, step_size=1, gamma=hp.lr_gamma)
```

7. These schedulers will decay the learning rate over time by multiplying the current learning rate by the `lr_gamma` hyperparameter, set to .999 for the example. After every training batch run on the generator, we step/decay the learning rate by calling the `step` function, as shown here:

```
scheduler_G.step()
scheduler_D.step()
```

8. Next, we will look at the discriminator loss calculation. Notice how we are using hinge loss for the adversarial fake and real loss and then computing total loss by attaching a gradient penalty.

```
loss_D_fake = F.relu(1 + fake_validity).mean()
loss_D_real = F.relu(1 - real_validity).mean()
loss_D = loss_D_real + loss_D_fake  + hp.lambda_gp * gradient_penalty
```

9. As Figure 7-11 demonstrates, this example can take a while to train and even produce something interesting. However, when it does, the output can be remarkably interesting and, in some ways, artistic.

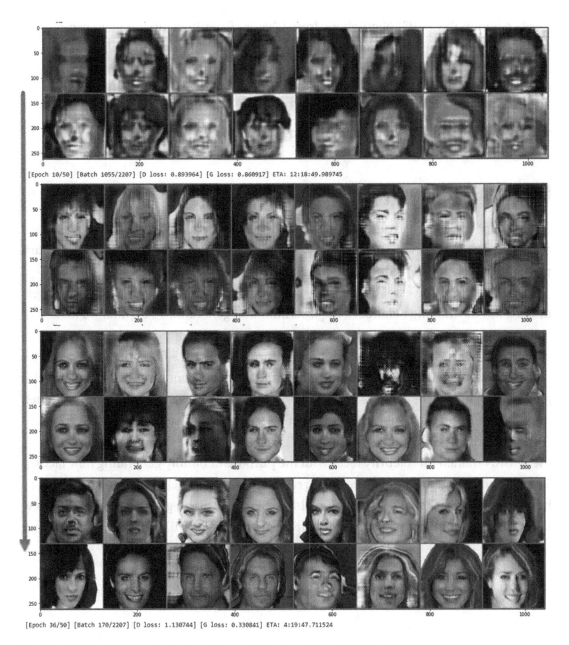

[Epoch 10/50] [Batch 1055/2207] [D loss: 0.893964] [G loss: 0.860917] ETA: 12:18:49.989745

[Epoch 36/50] [Batch 170/2207] [D loss: 1.130744] [G loss: 0.330841] ETA: 4:19:47.711524

Figure 7-11. Example output from training the celebrity SAGAN

This exercise and example may take a long time to train to get fantastic results, but the results you do get before then can be themselves mesmerizing. Take some time and watch the training of this example, and you will be rewarded by both some amusing and scary output.

As a bonus, the StarGAN notebook from Chapter 6 has been upgraded to use self-attention and spectral normalization. Be sure to also check out the GEN_7_Self-attention_STARGAN.ipynb notebook for more examples of using self-attention with spectral normalization.

That completes our whirlwind chapter on self-attention, the GAN that uses them, and the mechanisms that help support them.

Conclusion

In this chapter, we learned about attention and how to define it in terms of local/global and too soft or hard. Understanding attention allowed us not only to better define feature mappings or relationships but also to regenerate those same relationships.

However, using the extended feature mapping of attention also required us to better understand the training balance of the discriminator/generator. To do that, we looked at an important mathematical and abstract property of functions called Lipschitz continuity. This abstract term defines how uniform or smooth a function is.

We therefore learned that by constraining the Lipschitz continuity of a function using gradient penalty loss or spectral normalization, a GAN could be a more balanced trainer, thus allowing us to use better feature extraction strategies like self-attention.

Finally, in this chapter, we put everything together in first an implementation of the SAGAN trained on celebrity faces with the purpose of blindly generating new faces. Then improving on this model, we added residual network blocks for feature skipping an improvement we borrowed from the previous chapter, again to generate completely new faces, this time at higher resolution and ideally with more detail.

CHAPTER 8

Advanced Generators

The field of artificial intelligence and machine learning is growing by leaps and bounds daily with new forms of GANs/generators being developed. Through the course of this book, we have covered the history of those generators as they have progressed. We looked closely at the details and technical advances of their contribution to generation, as well as how they may have failed.

Our goal in this book has been to provide you with a good foundational knowledge to understand the nuances of generators and how to build them. Using a platform like Google Colab provides significantly easier access for students to learn complex algorithms without the agony of installing the dependencies themselves. This can go a long way into making generators and deep learning far more accessible to newcomers.

Unfortunately, with AI/ML advancing so rapidly, there comes a point of diminishing returns, where the code examples get so big and so many new variations or features are implemented that explaining a single new form of GAN could take the rest of this book. Therefore, for this and the next couple chapters, we are now going to take a more functional approach to using generators with packaged code.

In this chapter, we take our first look at using packaged open source code that can be quickly set up and either trained or used for actual generation. All the examples we will look at in this chapter are advanced versions of generators. And while we won't look at their code, we will take a high-level look at how they function. Since the code is all open source, interested readers can then advance to investigate the inner workings of that code themselves.

We will start the chapter by looking at the progressive GAN (ProGAN). A method develops a generator by progressively building up image resolutions through training. From there we will move up to the StyleGAN, a style transfer–inspired GAN that builds from ProGAN. We will look at StyleGAN2, or version 2, which can produce some excellent results trained on our old friend CelebA.

© Micheal Lanham 2021
M. Lanham, *Generating a New Reality*, https://doi.org/10.1007/978-1-4842-7092-9_8

Training advanced GANs, especially on a cloud notebook, is less than ideal, so in the final sections of the chapter we look at using other generator packages for generation. We will look first at DeOldify, a newly inspired generator referred to as a NoGAN that can colorize old photos and videos. Then we finish the chapter with another non-GAN implementation called ArtLine, which turns photographs into line art drawings.

This chapter is a blend of training advanced examples like ProGAN and StyleGAN2 to using cool generators like DeOldify and ArtLine. Compared to previous chapters, we will keep the code within a black box and look at only what is required to get the package running. Here is a summary of the main elements we will cover in this chapter:

- Progressively growing GANs

- Styling with StyleGAN version 2

- Using DeOldify, the new NoGAN

- Being artistic with ArtLine

If you have been playing along at home and following the examples throughout this book, you will surely be a deep learning and GAN training master by now. While we will continue to borrow from some of those skills in this chapter, we take a more functional approach to generation. Before that, though, we are going to look at the foundations of progressive GAN growing in the next section.

Progressively Growing GANs

In 2016, a team at NVIDIA wrote a paper titled "Progressive Growing of GANs for Improved Quality, Stability, and Variation." This paper outlined a procedure for progressively growing a GAN, starting with very low-resolution images to increasing resolution. A ProGAN may start with a 4×4-pixel image and work its way up to a 1024×1024 high-resolution image.

The concept of a ProGAN originated with trying to solve the problems we often encounter with convolution. As we have seen time and again throughout this book, the deeper convolutional layers become, the more noise they project. This was observed in various chapters where we explored deep CNN layers for feature extraction and reproduction.

We have seen several solutions to managing the deep convolutional problem, including batch normalization, UNets, ResNets, and self-attention with spectral normalization. While these have and are successful on their own, the concept of a ProGAN goes a step further in breaking down training into progressions.

After all, you couldn't necessarily expect to teach calculus to an elementary student, so why should we expect an untrained GAN to learn to generate a face from no past experience? As much as we feed thousands of faces in thousands of iterations to try to teach such models, there becomes an inefficiency in that form of training.

That inefficiency, as we now know, occurs when our models get too big and too deep with tens or hundreds of convolutional layers. But we need to create such big models if we are going to train models to generate high-resolution images. However, instead of building those models from the start, ProGAN builds a model incrementally through training.

Figure 8-1 shows the progression of training a ProGAN on facial data. The process works by first reducing the real training images to 4×4 pixels and feeding those into a GAN designed for that image size. After a certain period of training time, the GAN is grown by adding new layers for higher resolution, first 8×8 but then up to 16×16, 32×32, 64×64, and so on, to 1024×1024.

The sacrifice of progressively growing a GAN is the additional training time of building successive models over and over but at higher resolutions. Additionally, training such models requires extra data preparation and storage. These are not ideal requirements when using a cloud notebook like Colab, so we will stick with some simpler examples in the exercise.

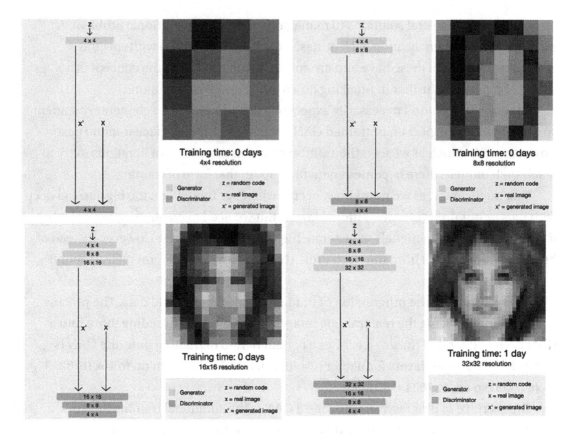

Figure 8-1. *ProGAN training progression*

Figure 8-2 shows the internal structure of a ProGAN for the generator and discriminator as proposed by the original authors. In the figure you can see the base blocks used to define the first progressions of the GAN from 4×4 to 16×16. As the model continues to build to whatever resolution is needed, new blocks are formed using the template shown, where k represents the desired pixel resolution of the final output image.

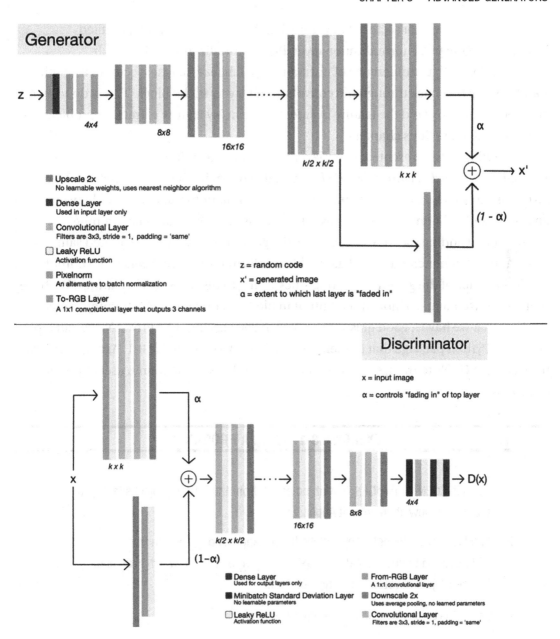

Figure 8-2. *Internals of ProGAN*

Within the generator pictured in Figure 8-2, there is a new layer type called Pixelnorm, or the pixel normalization layer. Pixel normalization is like batch normalization with the difference of normalizing each feature pixel vector to a length of 1 and passing that back through the convolutional layers in the block. This is done with another method to avoid the feature over feature or noisy feature extraction problems we encountered in previous chapters.

The authors also presented a new layer form for the discriminator as well called *minibatch standard deviation (MSD) layer*, again shown in Figure 8-2. This new layer form was introduced to add a means for the discriminator to track statistics between real and fake batches of images, using those statistics as an extra channel in training.

A result of adding the MSD layer forces the generator to vary its generated samples in a manner that matches the real data. This results in more varied generated output and less chance of the generator getting stuck trying to resolve specific features. We have seen the consequence of nonvaried output in plenty of previous GAN training exercises.

Now that we have a good grasp of the high-level workings of a ProGAN, we can look at using a Python package that implements such a GAN in Exercise 8-1. Again, these can be expensive GANs to train, so we will look at a simple example training set to just get our feet wet.

EXERCISE 8-1. USING A PROGAN

1. Open the GEN_8_ProGAN.ipynb notebook from the GitHub project site. If you are unsure how, then consult Appendix B.

2. The first cell in the notebook installs the package pro-gan-pth from https://github.com/akanimax/pro_gan_pytorch, which is an open source project devoted to this type of GAN:

   ```
   !pip install pro-gan-pth --quiet
   ```

3. The imports and other setup code in this example is trimmed down and should be simple enough to follow for most readers who have progressed this far in the book. We will cover a few of the more interesting sections starting with the function shown here:

   ```
   def check_output():
       print("rendering output loop - started")
   ```

```
folder = './samples'
while running:
   time.sleep(15)
   file = get_latest_file(folder)
   if file:
      clear_output()
      print(file)
      visualize_output(file,10,10)
```

4. This function is in place to render the output while the ProGAN code is training. To see continuous output in the notebook, we will use this function within a separate process external to the training cell. We do this so the output may be rendered in real time and visible while training.

5. Next, we scroll down to where the ProGAN is created and trained with the following:

```
pro_gan = pg.ConditionalProGAN(num_classes=10, depth=depth,
                latent_size=latent_size, device=device)
with io.capture_output() as captured:
    pro_gan.train(
        dataset=dataset,
        epochs=num_epochs,
        fade_in_percentage=fade_ins,
        batch_sizes=batch_sizes
        )
```

6. The default training implementation of this package is quite noisy and doesn't work well within a notebook. Therefore, we turn off the output using the function io.capture.output() to suppress the cell's output.

7. From here we can look at how this code is all run by instantiating a working additional process for the rendering and a worker thread for the training.

```
t1 = threading.Thread(target=train_gan)
p = multiprocessing.Process(target=check_output)
```

```
start = time.time()
p.start()
t1.start()

t1.join()
```

8. Again, this code is in place so that we may see the training output of the GAN as it progresses. Since the GAN is created in another thread called t1, we use t1.start and t1.join to start and wait for the thread to complete. Likewise, p is created for the process that will run the rendering loop. However, be aware that the rendering loop process will continue if you terminate the cell, so to stop the process, you need to do a runtime restart from the menu.

9. From the menu, select Run ➤ Run all to start the training and see the results as the ProGAN is trained.

As the previous exercise trained, you saw the model progress to various resolutions starting with 4×4 and working up. The final output of this example is not outstanding since it upscales to only three generations, but keep in mind that as the model improves, higher resolutions will continue to look better as training progresses.

You can, of course, play with other variations of this GAN from training other sample data to exploring different features of the Python package yourself. If you want to explore this GAN more, just consult the GitHub resource page for further documentation. The ProGAN is a good demonstration for the next big step in GANs we will look at in the next section.

Styling with StyleGAN Version 2

The original StyleGAN was first unleashed in another paper from the folks at NVIDIA titled "A Style-Based Generator Architecture for GANs" that extended their work of the ProGAN. This form of GAN was developed to extend the generation capabilities of the generator and in particular distinct feature generation.

What the authors found was that they could isolate feature extraction and replication into three distinct classes or granularity. This, in turn, projected into the architecture of the GAN itself; at lower/top layers, coarser features are extracted, with lower blocks identifying more detailed features. The authors defined the granularity of those feature extractions, as shown here:

- **Coarse (less than 82 pixels)**: Identified at the low details and would include features such as hair, face orientation, and size

- **Medium (from 162 to 322 pixels)**: Typically defined by features such as finer facial features, eyes closed, and mouth open

- **Fine (642 and up pixels)**: Tunes to the detailed features like eye, hair, and skin color

Based on this concept of extracting features at the layer level, StyleGAN extended the model by providing two new major enhancements. The first, a mapping network, provided the ability to map feature vector input to actual visible features. Second was the addition of style modules that converted those feature mappings into visible features.

Mapping Networks

Mapping networks provide a nontrial method of converting an encoded vector representation into visible features. These forms of network work similarly to the encoder portion of an autoencoder with the additional step of identifying visible features. The produced artifact of mapping encodings to features is called *feature entanglement*.

Figure 8-3 shows the mapping network and details the breakdown of the eight layers used to map features to vectors using layers and reduces to the same size as the input. Within the figure you can see how the result, called W, is then fed into a ProGAN, synthesis network for generation.

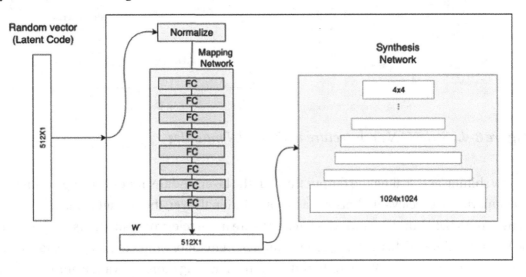

Figure 8-3. *Mapping network in StyleGAN*

Style Modules

Style modules, or adaptive instance normalization (AdaIN), take the vector output from the mapping network, W, into the model's generation layers. One module is added between each upsample and convolutional layer within each respective progression block.

Figure 8-4 shows the addition of the AdaIN modules and the process of using the encoded W from the mapping network as direct inputs to each module. In some respects, this is like residual networks where we skipped across layers by allowing inputs to bypass in some form. The AdaIN modules do not use the full input but rather a discovered mapping output as W; they also only provide scaling/normalization of the input.

Within the blowup diagram of Figure 8-4, you can see the inner workings of how the encoded vector is used to scale and normalize the output through the layers. That means AdaIN layers are not additive like ResNET but define a scaling normalization process that is applied to the outputs.

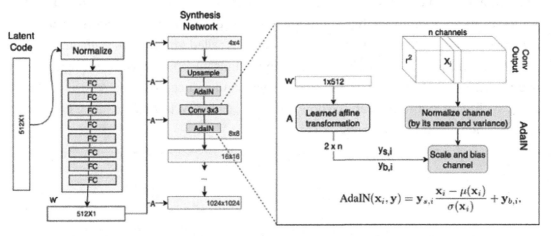

The generator's Adaptive Instance Normalization (AdaIN)

Figure 8-4. *StyleGAN architecture with AdaIN modules*

By shifting and scaling the output through the layers, AdaIN modules help promote the importance of relevant filters in each convolutional layer, thereby allowing the generator to better understand which features are more relevant than others. It may also be helpful to think of this as an internal reinforcement loop, where the mapping layers learn how to encode the W vector based on understanding more relevant features.

Removing Stochastic/Traditional Input

Since the mapping network only relies on reducing the stochastic input to an encoding, the authors decided to discard the need for randomness. Instead, they used constant inputs as the initial vector passed in. This dramatically reduced issues with entangled features by minimizing or controlling the random inputs into the generator.

Entangled features occur when disparate features may get inadvertently mapped together like the position of the hair. During our previous training exercise, you likely noticed this countless times, where feature-like hair strands may appear next to a face rather than on top.

Stochastic Variation (Noisy Inputs)

Having removed stochastic input, there became a need to add some form of variation into the models. Without variation, a model will become more specialized. The addition of noise allows the model to keep more generalized. What the authors found was they could introduce variation in outputs by adding random noise to each channel before the input is passed into the AdaIN layer.

The benefit of adding random noise at each style block allows for control of the finer details of the model generation, thus allowing for finer feature details such as freckles, facial hair, and wrinkles/dimples to be generated at random. Without the random noise, finer details would be obscured by more granular features.

Mixing Styles

One downside of using the W intermediate encoded vector as direct input to each AdaIN block is the tight correlation between model features. To break this correlation, two sets of inputs are chosen and passed through the mapping network, where each output is then randomly passed with a 50/50 chance to each AdaIN block.

While this doesn't benefit all forms of training data used on StyleGAN, it does have an upside effect with homogenous datasets like CelebA. What the authors discovered was that they could combine features from one generated image with a second generated image to produce a third new combined image.

An excellent video demonstrating this effect is on YouTube at `https://youtu.be/ kSLJriaOumA`. This video provides an excellent visual of how image features may be combined to produce new unique combined images.

Truncation of W

As we have seen through course of several chapters now, poorly represented areas in real data are not easily generated. For example, in our CelebA dataset there is only a small percentage of images with bald people. In previous generators, this caused the generation of bald people to be poor since the sample was poorly represented.

To accommodate this common defect in training, the authors implemented an averaging of the encoded vector W. A continuous average of W is maintained called *wavg*, and then the encoded vector W is transformed into a delta or difference from the average.

The thought here is that producing the best average image and working back by allowing the image to be modified with w_delta also provides for additional feature control. This in turn allows the model to be controlled just by altering the w_delta, not unlike our mapping across mean and variance with a variational autoencoder.

Hyperparameter Tuning

The last major element the authors of the StyleGAN contributed heavily to was spending hours tweaking and tuning the model's hyperparameters. They made significant improvements with respect to their previous efforts of the ProGAN. The hyperparameter values for learning rate, training epochs, and so on, have all been optimized in the code.

As a result, the StyleGAN and StyleGAN2 we will look at shortly produce incredible output with respect to faces. If you decide to use this model on your own datasets, you may need to spend time modifying the various hyperparameters accordingly.

Frechet Inception Distance

The output of generative models is measured on a scale called the Frechet Inception Distance (FID). This distance is measured based on comparing the activations within pretrained image classification networks on the real and output images. Low scores represent better generated model outputs.

In Figure 8-5 there is an FID comparison of StyleGAN based on top of ProGAN at the top with the second version StyleGAN2 below it. FFHQ stands for Flikr-Faces HQ, which is another facial dataset used to train generators. You can see the results go from an FID score of 7–8 in the first table starting with ProGAN and reduce to just below 3 for the StyleGAN2.

These are impressive results, and if you want an interesting visual showing you just how well StyleGAN2 performs, visit `whichfaceisreal.com` for an excellent visual of how close these generated faces look.

Method	CelebA-HQ	FFHQ
A Baseline Progressive GAN [30]	7.79	8.04
B + Tuning (incl. bilinear up/down)	6.11	5.25
C + Add mapping and styles	5.34	4.85
D + Remove traditional input	5.07	4.88
E + Add noise inputs	**5.06**	4.42
F + Mixing regularization	5.17	**4.40**

FID results reported in the first edition of StyleGAN, "A Style-Based Generator Architecture for Generative Adversarial Networks" authored by Tero Karras, Samuli Laine, and Timo Aila. Note the FID scores on the right on the FFHQ dataset to compare with the StyleGAN2 resutls below.

Configuration	FFHQ, 1024×1024			
	FID	Path length	Precision	Recall
A Baseline StyleGAN [24]	4.40	195.9	**0.721**	0.399
B + Weight demodulation	4.39	173.8	0.702	0.425
C + Lazy regularization	4.38	167.2	0.719	0.427
D + Path length regularization	4.34	139.2	0.715	0.418
E + No growing, new G & D arch.	3.31	**116.7**	0.705	0.449
F + Large networks	**2.84**	129.4	0.689	**0.492**

FID results reported in the second edition of StyleGAN, "Analyzing and Improving the Image Quality of StyleGAN" authored by Tero Karras, Samuli Laine, Miika Aittala, Janne Hellsten, Jaakko Lehtinen, and Timo Aila. Note the FID scores on the far left for the sake of comparison with StyleGAN1.

Figure 8-5. *FID score comparison of StyleGAN versus StyleGAN2*

While we won't have time to review the StyleGAN in more detail since the amount of code to cover all those features is substantial, we will instead progress to the gold standard in facial generation now. That standard is currently set by StyleGAN2 and something we will cover in more detail in the next section.

StyleGAN2

While the team at NVIDIA could have been certainly happy to stop with StyleGAN, they continued to push the limits of what is possible and as result produced StyleGAN2. This version further reduced the FID score from 4.40 to 2.84 and produced some of the most realistic faces yet.

Like the previous section, when we looked at the StyleGAN, we will review each of the feature groups that provided increases in FID score with respect to Figure 8-5. We start with the first feature weight demodulation in the next section.

Weight Demodulation

The AdaIN layers are a derivation from an earlier concept called *neural-style transfer* where styles could be captured and transferred. However, what the authors of StyleGAN2 found was that visual artifacts such as water droplets or smudges could be carried through images by the reinforcement of the styles.

Instead, what they discovered is that the AdaIN layers could be moved to inside the convolutional layers themselves and not used as direct inputs but rather baked in as normalization.

The benefit of baking in the style normalization to the convolutional layers themselves provided for parallelization of the computations, thus allowing for models to train upward of 40 percent faster and further removing visual artifacts like droplets or smudges.

Path Length Regularization

This introduces a new normalization term to the loss to better smooth or make uniform the latent space. We have seen throughout this book the importance of normalizing or making uniform the latent space in generative models. It has been a critical concept in understanding how to producing models that can generate consistent output.

Smoothing the latent space provides for the mapping of images more readily into and out of the space in known projections allows for more controlled generation of images but also provides for projection of images back into latent space encodings. Understanding this relationship from latent space to generated images allows for images to be generated across a path of latent space.

The capability to smooth and understand the latent space provides further applications from animation to age progression and more. These have been demonstrated successfully using StyleGAN2 to animate across styles and features.

Lazy Regularization

Applying path length regularization can be a computationally expensive process that doesn't necessarily derive benefit through every training iteration. As such, in StyleGAN2, PLR is applied only every nth number of iterations where n is typically set to 16 but may be varied.

No Growing

As we saw, progressively growing GANs were successful in building up large-scale images from 4×4 to 1024×1024 pixels in size. However, what typically happens with a ProGAN is a strong location preference for certain features such as a nose, eyes, and mouth.

To bypass these issues yet still allow for the progressive development of the generator model, we turn to another recent paper: "Multi-Scale Gradients for Generative Adversarial Networks" by Animesh Karnewar and Oliver Wang. This paper introduced the idea of multiple scale gradient using a single architecture.

Figure 8-6 shows the architecture of the multiple scale gradient GAN (MSG-GAN), not to be confused with the food additive. In the MSG-GAN, the model is designed to progressively build generated output based on the same progressions we see in a ProGAN.

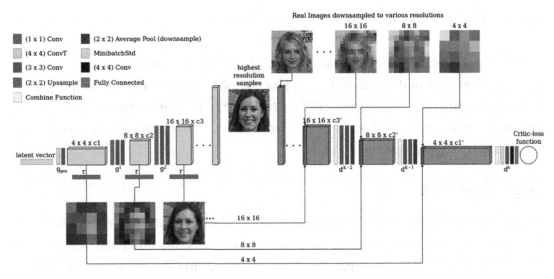

Figure 8-6. *MSG-GAN architecture*

Authors of StyleGAN took this progression model and implemented it with input/output skips not unlike what we have seen in a ResNet. Figure 8-7 demonstrates a comparison of the architecture differences between the three models MSG-GAN, input/output skipping, and ResNet.

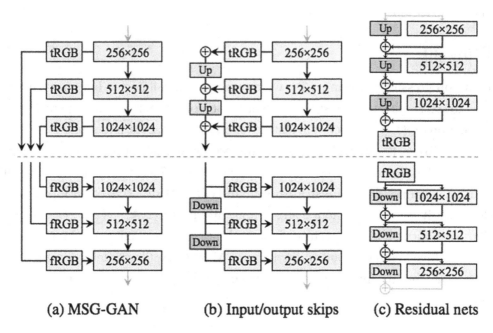

(a) MSG-GAN   (b) Input/output skips   (c) Residual nets

Figure 8-7. *Comparison of progressive architectures*

The skip architecture used in StyleGAN2 can perform the same progressive growing of feature granularity as is done in ProGAN, allowing the model to focus more on the higher progressions (those at 1024×1024 than through normal progressive development). When scaled up using large networks, this enhancement was further magnified, as we will see next.

Large Networks

Through our exploration of ResNet, we saw how a typical shallow model with 10 or fewer layers of convolution could be increased to generators with more than 100, all through the magic and use of skipped connections. Not unlike ResNet, StyleGAN2 can also benefit from exceptionally large networks through the adaptation of input/output skipping.

Figure 8-8 compares the respective feature contribution between StyleGAN and StyleGAN2. Remember that StyleGAN uses a basic progressive growing architecture, whereas StyleGAN2 uses MSG input/output skipping to allow the model to focus more on the finer details we typically see generated at the higher levels like 1024×1024. The x-axis represents a level of progression, and the y-axis denotes the percentage of accuracy when generating an output of the preferred size.

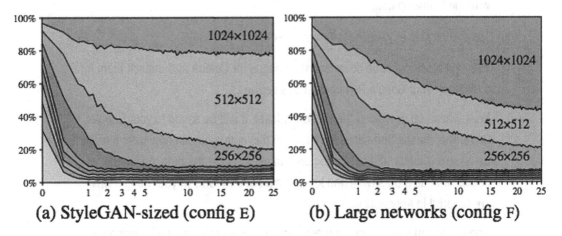

(a) StyleGAN-sized (config E)   **(b) Large networks (config F)**

Figure 8-8.   Comparison of feature contribution to respective outputs

By building large networks, the model generator can place more emphasis features developed from the higher layers. As Figure 8-9 shows, the 1024×1024 features are far more emphasized in the large networks as opposed to the regular StyleGAN.

Now that we understand all the features of the StyleGAN and StyleGAN2, we can move on to training a version using an already established package. In Exercise 8-2, we will train a PyTorch version of StyleGAN2 on our old friend the CelabA dataset. Let's jump into the next exercise and train a StyleGAN2 to do our bidding.

EXERCISE 8-2. TRAINING STYLEGAN2

---

1.   Open the GEN_8_StyleGAN2.ipynb notebook from the GitHub project site. If you are unsure how, then consult Appendix B.

2. This model can take a substantial amount of time to train, but the results are worth it, and likewise, the saved models generated are also worth keeping. With that in mind, for this notebook we will use the features discussed in Appendix C that allow us to connect to our Google Drive and save the models permanently. The top cells of the notebook provide for the connection to your Google Drive.

3. Next, we will install the PyPi package `stylegan2_pytorch` for the StyleGAN2 with the following code:

    ```
    !pip install stylegan2_pytorch --quiet
    ```

4. The next blocks of code download the images for CelebA and unpack them into a folder on your Google Drive called `stylegan2`.

5. Now, when you download the CelebA dataset, it will be saved to your Google Drive. This means that successive runs of this notebook can now refer directly to the saved folder and not be required to redownload the data. Again, the details of setting up saving and loading data and models to your Google Drive are covered in Appendix C.

6. Finally, we can simply run the model against the save image folder using the following code. Notice the use of the $ before the variable `image_folder`. We use the $ to substitute variables from the Python code to the shell.

    ```
    !stylegan2_pytorch --data $image_folder
    ```

7. When you run the last cell, the model will be set up and begin ingesting real images from the saved image folder. While running the code, output is saved into your Google Drive within the `stylegan2` folder in subfolders called `results`. Along with the generated output, models of the current training spot are also saved into another folder called `models`. Refer to Appendix C on how to load/save models.

8. You can view the generated output at any time by opening the files folder using the left folder icon. Then drill down to the `gdrive/MyDrive/stylegan2/results/default` folder, as shown in Figure 8-9a.

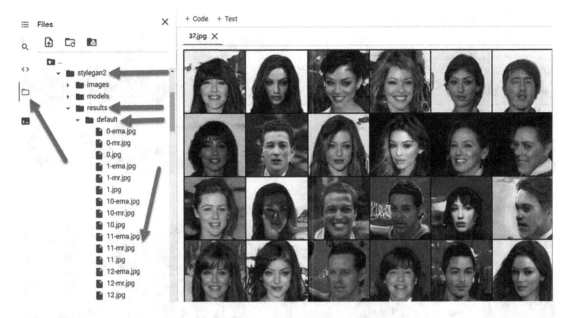

Figure 8-9a. *Examining generated output while training*

In Figure 8-9a, you can see the results of 10 percent training. Notice the key differences in how well the faces are generated in comparison to our previous examples. There are still unpleasing visual artifacts in some images, but overall, the details in the images and faces are exquisite.

StyleGAN and StyleGAN2 really took the overall generation model and added several features to provide for excellent feature extraction and isolation in images, so much so that it is difficult for even us humans to determine which is fake or real now. We will look in Chapter 10 at methods for determining fake images and deep fakes.

Included with the previous exercise are two other datasets, `cars_all` and `foods`. The `cars_all` set is 60,000 photos of newer models taken in various perspectives from the exterior and interior. The `foods` dataset contains 80,000 photos of various cuisines on plates, bowls, or other serving media. Figure 8-9b shows examples of both datasets trained to 100 percent completion, which takes about two to three days on Colab and may take several restarts.

cars_all dataset

foods dataset

Figure 8-9b. *Output of StyleGAN2 trained to 100 percent on cars_all and foods datasets*

___

Our last experiment with StyleGAN2 used the CelebA dataset as a reference and comparison to our previous work with various GANs. Applying the StyleGAN2 to other datasets has produced remarkably interesting results from fake anime/cartoon characters to age progression and more. The `stylegan2_pytorch` package is a good base for any future generation work you may consider doing. For now, though, we will continue moving on in this chapter to the NoGAN in the next section.

DeOldify and the New NoGAN

As we saw from the start of this chapter, generative modeling is advancing quickly and with it the requirements for entry. Many may feel that building new and better generators is beyond the scope of anyone outside academia or a dedicated research lab. Fortunately, this is far from the case, and in fact, new advances are made every day outside research institutions.

One such individual who has made their mark is Jason Antic, a self-described programmer who took an advanced course in AI from Fast.ai. He became so enthralled with AI after taking the course he reduced his hours at work and dedicated all his time to building new generators. Antic would go on to produce a notable GAN called DeOldify that is capable of colorizing and enhancing old photos.

The first version of DeOldify was developed from a progressive self-attention GAN and had options for colorizing from artistic to normal. DeOldify quickly became a huge hit, and many people took notice how a simple programmer could accomplish so much in a little amount of time. Antic now works on the project full-time and has teamed up with his old instructors at Fast.ai.

From his work on DeOldify using typical GAN techniques, Antic would eventually conclude that regular training was flawed. What he found was that he could get better results by training the generator and discriminator separately for the bulk of training and then bring them together during the late stages of model development.

The actual process works by first training the generator to create images and using the feature loss for training. After the generator has achieved a desired minimum feature loss, the generated images are trained against the discriminator as a binary classifier, real or fake. Then when the critic is trained sufficiently, both models are brought together using a typical GAN training method.

Antic named this new form of GAN a NoGAN since the bulk of training was done in compartments. From this form of training, Antic also discovered an interesting characteristic in the way models learned. What he observed is at some time in training there is an inflection point between great output and that with distorted features. To find this inflection point, he needed to be rigorous in testing multiple model variations and training points.

By observing this inflection point and determining when models were optimal, Antic would then go back and retrain the opposite model. The results of his work are quite impressive, as we will see in two exercises for this section. In Exercise 8-3, we will use the DeOldify packaged model to colorize some old and historical black-and-white photos.

EXERCISE 8-3. USING DEOLDIFY TO COLORIZE IMAGES

1. Open the GEN_8_DeOldify_Image.ipynb notebook from the GitHub project site. If you are unsure how, then consult Appendix B.

2. This notebook needs to download, install, and restart before we can use the magic of DeOldify. That means you will need to run each of the top cells individually until you get to the required installation for DeOldify, as shown here:

```
!pip install -r colab_requirements.txt
```

3. Since we are pulling DeOldify directly from its GitHub repo, we need to install the various requirements for the package. After installing the requirements in the last block of code, you may need to reset the notebook's runtime. From the menu, select Runtime ➤ Restart runtime.

4. After installing DeOldify, you can move on to the familiar blocks of code to download the test/training images we have used throughout this book and won't need to review here. In this example, we are using historical black-and-white images, but you could also use any of your own images.

5. We are using DeOldify with Python code and will import the various required libraries in the Import section. Notice the use of the fastai import as this is a primary library used for the package.

```
import fastai
from deoldify.visualize import *
import warnings
warnings.filterwarnings("ignore", category=UserWarning,
message=".*?Your .*? set is empty.*?")
```

6. Next, we will jump down and look at where we download the pretrained models and a watermark. The watermark is used to denote that the images are generated from AI.

```
!mkdir 'models'
!wget https://data.deepai.org/deoldify/ColorizeArtistic_gen.pth -O ./
models/ColorizeArtistic_gen.pth
```

```
!wget https://media.githubusercontent.com/media/jantic/DeOldify/
master/resource_images/watermark.png -O ./resource_images/
watermark.png
```

7. We can then instantiate a `colorizer` from the package using just a single line of code, like so:

```
colorizer = get_image_colorizer(artistic=True)
```

8. With the `colorizer` created, we can move on to using it in a Colab form with the following code:

```
#@title COLORIZE IMAGES  { run: "auto" }
import glob
from PIL import Image
import ipywidgets as widgets
from IPython.display import display
from IPython.display import clear_output

files = sorted(glob.glob("%s/*.jpg" % image_folder))
file_idx = 27 #@param {type:"slider", min:0, max:35, step:1}

show_image_in_notebook(files[file_idx])
image = colorizer.plot_transformed_image(files[file_idx])
show_image_in_notebook(image)
```

9. Using the form, you can then slide through the various images and see what each looks like colorized. Figure 8-10 shows the output of the original image beside the DeOldify colorized version.

Figure 8-10. *Colorized and enhanced image*

Unfortunately, you won't be able to see the full effect of colorization if you are reading this book in print. To see the full capabilities and enjoyment of colorizing photos quickly, you will need to run the notebook exercise. The results are quite impressive, and you will surely be a hit with your older relatives asking you to colorize old photos.

You will notice that a watermark is placed in the bottom-right corner of the image. In Figure 8-10 this is obscured by a previous watermark. The art palette watermark that you can clearly see on the other images denotes that the image has been generated from AI. There is a growing consensus of AI developers/researchers that all generated content should bear this watermark.

Colorizing and Enhancing Video

DeOldify is capable of colorizing not only images but video as well. The specific model variation used to colorize video is called *simple video* and uses a few improvements on its own. While the video model is based on SAGAN/NoGAN training, it is improved by optimizing across frames to prevent flicker-free output.

In this next exercise, we set up the DeOldify video model to process older movies that we want to colorize. For the most part, this example will be like the image (artistic) model we used to colorize in our previous exercise. Colorizing video can be quite interesting to see how frames are interpolated across output, as we will see in Exercise 8-4.

EXERCISE 8-4. USING DEOLDIFY TO COLORIZE VIDEO

---

1. Open the GEN_8_DeOldify_Video.ipynb notebook from the GitHub project site. If you are unsure how, then consult Appendix B.

2. This notebook again needs to download, install, and restart before we can use the magic of DeOldify. That means you may need to restart the notebook's runtime after installation of the requirements.txt file, as shown in the following code block:

    ```
    !pip install -r colab_requirements.txt
    ```

3. Remember, you can restart the runtime from the menu by selecting Runtime ➤ Restart runtime. Note that runtime restart is different than a factory reset of the notebook. We typically only use factory resets when we want everything, including files and installations, to be reset.

4. The next few blocks of code are all like our previous exercise and other examples. Notice how this time we are loading a video-bw for video black and white dataset of old movies. Several of these clips are from the silent era of film; others may have audio. DeOldify strips out the audio from the file and replaces it after colorization, thus allowing movies with sound to be colorized as well.

5. Next, we can create the video colorizer with the following block of code:

    ```
    colorizer = get_video_colorizer()
    ```

6.   The key difference is the call to the helper function to create the video colorizer.

7.   From there, we can move down to the last block of code and see how the colorizer is again set up in a Colab form for easy use:

```
#@title IMAGE SELECTION  { run: "auto" }
import glob
import ipywidgets as widgets
from IPython.display import display
from IPython.display import clear_output

files = sorted(glob.glob("%s/*.mp4" % video_folder))
file_idx = 4 #@param {type:"slider", min:0, max:12, step:1}

show_video_in_notebook(files[file_idx])
video = colorizer.colorize_from_file_name(files[file_idx])
show_video_in_notebook(video)
```

8.   Again, after all the previous notebook cells have run, you can use the form to scroll through the collection of videos and visualize the output.

9.   Feel free to use your own selection of older videos and colorize/enhance them with DeOldify.

Figure 8-11 shows an example of a simple frame of video output. In the sample frame, we can see some artifacts where the ladies dress and horse's leg are not correctly colorized. When this type of similar discoloration occurred with skin, Antic would name these defects *zombie skin*.

Figure 8-11. *Example frame of colorized video*

Zombie skin describes gray or incorrectly colorized skin often as the result of poor feature extraction. In Figure 8-12 we can observe this zombie-like skin effect happening to the horse's leg, which appears to be gray. With video this effect is more pronounced as the scene lighting may quickly vary and cause these various artifacts. Again, the best way to visualize these results is to run the notebook and scroll through the various sample videos.

DeOldify is a great example of how generative modeling is becoming accessible to those outside research institutions. It demonstrates how a keen programmer can extend a model on his own to produce some significant and exciting results. We can certainly look forward to more projects like DeOldify in the future, and in the next section we look at another similar initiative called ArtLine.

Being Artistic with ArtLine

ArtLine was another project developed to turn photos into line art drawings and convert them to cartoon versions. The project follows many of the same discoveries that Antic used in building DeOldify. Both projects are based on many of the improvements/ utilities developed by the team at Fast.ai.

Technically, ArtLine is like DeOldify in that it is based on a self-attention GAN that is progressively resized to achieve a desired resolution. It is likewise trained as a NoGAN where the initial generator feature loss is determined by a pretrained VGG model much like we did in examples in Chapter 6. Unlike DeOldify, however, ArtLine does not use a discriminator for fine-tuning feature extraction, so it is never trained as a GAN.

As we will see, the results of ArtLine work well with portraits and faces and less so with other types of photos. It can, however, be interesting to see how this form of NoGAN works across a variety of images, as we will see in Exercise 8-5. We will again use the project's source code pulled directly from the repo and run it in a notebook.

EXERCISE 8-5. GETTING ARTISTIC WITH ARTLINE

---

1. Open the GEN_8_ArtLine.ipynb notebook from the GitHub project site. If you are unsure how, then consult Appendix B.

2.   This notebook again needs to download, install, and restart before we can be
     artistic with ArtLine. As such, you will likely need to restart the runtime after
     installing the `requirements.txt` file, as shown in the following code block:

```
!pip install -r colab_requirements.txt
```

3.   If you encounter an error further down in the notebook, make sure that you
     have restarted the runtime by selecting Runtime ➤ Restart runtime from the
     menu.

4.   The `imports` block for this notebook does a wide variety of installs for various
     components. Unlike DeOldify, this project has some requirements to describe
     the feature loss in a class.

5.   Scroll down to the `FeatureLoss` class and see how the model determines
     the loss between features. This class is used by the ArtLine model to evaluate
     feature training within the model itself.

```
class FeatureLoss(nn.Module):
    def __init__(self, m_feat, layer_ids, layer_wgts):
        super().__init__()
        self.m_feat = m_feat
        self.loss_features = [self.m_feat[i] for i in layer_ids]
        self.hooks = hook_outputs(self.loss_features, detach=False)
        self.wgts = layer_wgts
        self.metric_names = ['pixel',] + [f'feat_{i}' for i in
        range(len(layer_ids))
            ] + [f'gram_{i}' for i in range(len(layer_ids))]

    def make_features(self, x, clone=False):
        self.m_feat(x)
        return [(o.clone() if clone else o) for o in self.hooks.
        stored]

    def forward(self, input, target):
        out_feat = self.make_features(target, clone=True)
        in_feat = self.make_features(input)
        self.feat_losses = [base_loss(input,target)]
        self.feat_losses += [base_loss(f_in, f_out)*w
                        for f_in, f_out, w in zip(in_feat, out_
                        feat, self.wgts)]
```

```
        self.feat_losses += [base_loss(gram_matrix(f_in), gram_
        matrix(f_out))*w**2 * 5e3
                        for f_in, f_out, w in zip(in_feat, out_
                        feat, self.wgts)]
        self.metrics = dict(zip(self.metric_names, self.feat_losses))
        return sum(self.feat_losses)

    def __del__(self): self.hooks.remove()
```

6. Inside this class you can tweak one hyperparameter that scales the feature loss
    back into the model. The constant is exposed as 5e3, and you may alter this
    value to determine the effect it has on the final output.

7. Next, we can see how the pretrained ArtLine model is downloaded with the
    following code:

```
MODEL_URL = "https://www.dropbox.com/s/starqc9qd2e1lg1/ArtLine_650.
pkl?dl=1"
urllib.request.urlretrieve(MODEL_URL, "ArtLine_650.pkl")
path = Path(".")
learn=load_learner(path, 'ArtLine_650.pkl')
```

8. To use the model, we have again set up a form to allow you to scroll through
    the various images. For this example, we have two datasets, the `historic-bw`
    one we used in the previous exercise and a new one of interesting photos.

```
#@title ARTLINE IMAGES { run: "auto" }
import glob

files = sorted(glob.glob("%s/*.jpg" % image_folder))
file_idx = 7 #@param {type:"slider", min:0, max:25, step:1}

img = PIL.Image.open(files[file_idx]).convert("RGB")
img_t = T.ToTensor()(img)
img_fast = Image(img_t)
show_image(img_fast, figsize=(8,8), interpolation='nearest');

p,img_hr,b = learn.predict(img_fast)
Image(img_hr).show(figsize=(8,8))
```

9. Running the code will produce output shown in Figure 8-12, which is the same old photo we colorized with DeOldify. If you run through the various other photos from either dataset, you may notice the higher-contrast images from the historic set translate better.

Figure 8-12. *Example output from ArtLine*

ArtLine has another variation of its model that allows you to convert images to colorized cartoons. This model works quite well and can be set up in a similar fashion to the exercise we just worked with. We leave it up to you to pursue this on your own and enjoy the fun of converting images to cartoons.

Projects like ArtLine and DeOldify are recent initiatives brought about by developers interested in developing generative modeling for various applications, from creating art to coloring and enhancing old photos. How these projects mature and impact various communities remains to be seen, but it is also likely many other developers will follow suit.

Conclusion

As we saw in this chapter, generative modeling extends beyond the typical GAN architecture. It has grown into other variations from models that use progressive architectures to those that use no GAN training at all. How this whole new area of application development evolves in the short- and long-term remains to be seen.

One thing is for sure, however, which is that these forms of enhanced and impressive generators are only becoming more accessible. In the future, this should certainly have an impact on a wide variety of applications across multiple industries. The main hurdle we need to overcome now is the application of generative modeling to other sources of domain data.

However, it is also important as generative modelers that we understand how we can put this technology to use. In the next chapter, we look to apply all our learnings of generative modeling into the scariest application for many deep fakes.

## CHAPTER 9

Deepfakes and Face Swapping

There is likely no more controversial application of generative modeling than creating fake faces or applying the technology to swap faces. This technique is colloquially known as *deepfakes* and has been the basis for fake news and all forms of various related conspiracy theories. Many, from fear of understanding, see this technology as providing no value and as unethical, which also gives generative modeling a bad image.

Deepfakes have been used to create various memes and other YouTube videos of celebrities in famous roles with their face swapped out for the purpose of humor. Deepfakes have also been used to puppet political figures into various comedic and not so funny demonstrations.

Of course, there is a darker side to deepfakes where various pornographic images/videos have the nameless, often female, star's face swapped with a celebrity. This is by far the most unethical application of deepfakes since the celebrity's face is used without consent and placed in extremely compromising situations.

Yet, the power of deepfakes is also beginning to infiltrate Hollywood by providing new visual effects capabilities, where older or younger stars may have their appearance modified with deepfakes or more generally generative modeling. The current most common application of this technology is face aging or de-aging, which has been used in many blockbuster movies.

It remains to be seen how the application of deepfakes is used/abused, but what is demonstrated is how powerful generative modeling will have an impact on our future lives. For instance, it may be commonplace to swap faces or other forms in media. Imagine being able to swap out a star in a movie because you prefer another actor or, better yet, changing the whole style of the film. As we have learned through this book, swapping or generating anything is possible, given enough data and patience.

© Micheal Lanham 2021
M. Lanham, *Generating a New Reality*, https://doi.org/10.1007/978-1-4842-7092-9_9

We couldn't produce a comprehensive book about generative modeling without demonstrating the application of swapping faces. Understanding the how and what of deepfakes will allow you to theorize or possibly construct other forms of swapping. It also gives you the tools to promote the technology in an ethical and friendly manner.

In this chapter, we look at the application of face swapping (FS) and how it may be applied to photos, videos, and your personal webcam. We will look at a couple of tools that can be used to perform face swapping and go through the entire workflow to produce a deepfake video.

Then we will progress to training a face swapping model so that it can replace a subject's face with another. After that, we will use the software to convert the faces in a subject deepfake video of your choosing. We finish off the chapter with putting the whole process together and building a short deepfake clip of your own.

Face swapping can take a bit of work to get good results, but much like anything worth doing, the output can be rewarding and provide a sense of accomplishment. As we will see, it also demonstrates several aspects of applying generative modeling to a problem. It is an excellent way for us to start wrapping up this book. Here are the main points we will cover:

- Introducing the tools for face swapping

- Gathering the swapping data

- Understanding the deepfake workflow

- Training a face swapping model

- Creating a deepfake video

Unlike previous chapters, we will not strictly work with Google Colab for all the exercises. Instead, we will use a couple of GUI tools that are designed for the desktop. Not only will this make understanding the FS workflow easier, but it will provide you with a better long-term working environment. While the packages we use in this chapter are supported to run on Colab, it has been suggested that Google may prevent these types of apps from running freely in the future due to misuse.

Keep in mind that FS provides many opportunities for abuse, and our purpose for demonstrating how to build deepfakes is to promote your future ethical use. Many applications and operating systems are currently using facial recognition as a form of identification, and FS provides a potential security risk. As we will see in the next section, this technology is just getting started but already provides numerous sophisticated tools.

Introducing the Tools for Face Swapping

You may be surprised to learn that deepfakes and face swapping originated as far back as the 1990s when digital video manipulation first became possible. Since then, the technology has flourished in academia, film production, and most recently amateur communities made possible by commercialization and open sourcing.

There are a wide variety of techniques and applications that may perform some form of face swapping, from full digital video software that uses filters and other more artistic-driven techniques to those that apply generative modeling with deep learning. For our purposes, we are going to stick with community-driven projects that use or apply deep learning in some facet.

The following list shows the primary contenders we may consider for performing FS and producing deepfakes. This list is by no means comprehensive and will likely continue to grow in the coming years. For now, though it represents some of the more popular community-driven methods, we can use the following:

- **Deep Fake Lab** (`https://github.com/iperov/DeepFaceLab`): This is promoted as the top FS and deepfakes tool suggesting that 90 percent of content developed uses this tool. There are GUI and command-line versions of this tool that provide considerable options for managing the FS workflow. For that reason, it has become the most popular with several online videos demonstrating its use. It is also the more controversial tools since it hails from Russia and often avoids mentioning or promoting proper ethics.

  - *Base technology*: Face swapping/mapping is performed with various CNN autoencoders. Developed with TensorFlow.

- **Faceswap** (`https://github.com/deepfakes/faceswap#overview`): This less often used tool provides GUI and command-line versions for undertaking the FS workflow. It provides plenty of options and various plugins to control the workflow but is not as sophisticated as DFL. What makes this tool stand out is the ease of use with the GUI, including the installation, and a direct emphasis on ethical concerns when using it. For those reasons, it will be our chosen tool to demonstrate in this chapter.

  - *Base technology*: Various forms of CNN Aautoencoder models used for face-to-face translation. Developed with TensorFlow.

257

- **Faceswap-GAN** (https://github.com/shaoanlu/faceswap-GAN):
  This is an older, less maintained tool that is a great demonstration of
  doing FS with a GAN combined with a CNN autoencoder similar to
  the more popular tools mentioned earlier.

- **CelebAMask-HQ** (https://github.com/switchablenorms/
  CelebAMask-HQ): This is based on a MaskGAN, which like a
  CycleGAN is a model that provides for editing and extracting features
  from faces. Masking and editing features are only one part of the FS
  workflow, as we will see. This makes this tool not able to perform full
  FS but could be used to enhance other of the other tools.

- **StarGAN version 2** (https://github.com/clovaai/stargan-v2):
  This is a model extended from a StarGAN that can be used to alter or
  swap faces to match various learned attributes. This tool can be used
  to perform full face swapping or just swap out attributes like we did in
  Chapter 6 when we used version 1 of a StarGAN.

- **FSGAN** (https://github.com/YuvalNirkin/fsgan): This is
  described as a subject-agnostic face swapping and reenactment tool.
  This tool is one of the few developed with PyTorch and does provide
  a Google Colab version as well. It performs decent face swapping but
  requires a keen knowledge of the FS process. This is a good, advanced
  tool for anyone looking to upgrade their skills to using a GAN for FS.

One thing to notice from the previous list is that the two main applications used
to perform FS (DFL and face swap) use somewhat low-tech approaches with CNN
autoencoders. As we saw in the earlier chapters, autoencoders can produce effective
results and can be trained relatively quickly. They also have fewer hyperparameters to
tune, which makes them overall more robust for amateur usage.

Figure 9-1 shows the underlying architecture for the Faceswap tool as it is used for
training and then actual conversion. In the tool, we can see that the model encoder takes
two input images labeled face A and B. Face A represents that target face you want to
swap out, and B is the face you want to replace with.

Within this architecture, you can see that the encoder splits to two decoders, one
for each set of faces. This configuration is like our earlier explorations of image to
image models using paired and unpaired translations. With Faceswap, we use unpaired
translations to teach the model the translation required to map one face to another.

Then, as shown in Figure 9-1, when it comes time to convert, only the respective input face (A) needs to be fed into and then converted using the decoder for B. The result is the converted B face that can then be swapped for the original face.

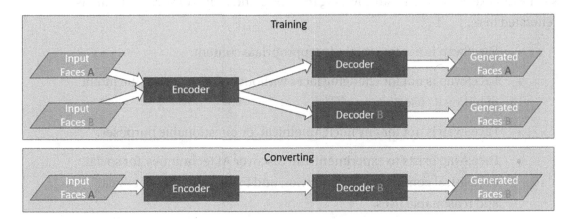

Figure 9-1. *Faceswap encoder-double decoder architecture*

Figure 9-2 shows the results of using Faceswap to swap Julia Roberts' face with Scarlet Johannsen's. Results were obtained using the Villian model, which is a third-party plugin often used to produce high-quality deepfakes. Figure 9-2 shows a comparison of the before and after of swapping the star's face in the movie *Pretty Woman*.

Figure 9-2. *Face swapping Julia Roberts for Scarlet Johannsen in the film Pretty Woman*

We will learn how to create the deepfake shown in Figure 9-2 through various exercises later in this chapter. For now, though, we want to reiterate the importance of keeping our ethics at the forefront when creating such content. As such, the Faceswap site has an excellent manifesto covering the usage and ethics of such tools and is reiterated here:

- FaceSwap is not for creating inappropriate content.

- FaceSwap is not for changing faces without consent or with the intent of hiding its use.

- FaceSwap is not for any illicit, unethical, or questionable purposes.

- FaceSwap exists to experiment and discover AI techniques, for social or political commentary, for movies, and for any number of ethical and reasonable uses.

Always try to keep these guidelines in mind when using this technology. There exist numerous ways for people's reputations to be harmed through abuse that could result in serious legal consequences. That aside, your abuse of deepfakes could potentially harm future employment opportunities or hopes for working in the AI generative modeling workspace. Be careful about what you produce and how you present it.

Provided you respect this technology, your subject's face swapping can be an entertaining pursuit that you could get lost in for days, weeks, or months. It can be amusing to swap celebrities in movies or images to see the what-if possibilities. There are plenty of excellent examples online, good and bad, featuring humorous and respectable use of this technology.

Whether you use this technology for personal use or create your own online YouTube channel, the FS workflow is common across toolsets. We will start in the next section looking at how we can gather data for the purposes of performing FS workflows.

Gathering the Swapping Data

Gathering required data for any AI/ML task, be it generative modeling or image classification, can be an undertaking in and of itself. Fortunately, there are numerous example datasets online that can be used for various tasks, as we have demonstrated throughout this book. However, at some point, there comes a need to gather your own data for the projects you want to work on.

Therefore, in this section, we will look at a few tools developed on Colab that can help gather data we need for proper face swapping and deepfakes generation. These tools rely on third-party community packages that are always in a state of flux and change. While it is hoped that these tools continue to function well into the future, you may need to look for alternatives.

Tools like YouTube-Downloader, which we will employ later in this section, are often plagued with controversy. These downloaders allow us to download videos directly from YouTube. However, YouTube is not so keen on allowing users to download videos without consent, so the tools may often alter its API to break this software. GitHub has even gone so far as to unsuccessfully try to ban such tools from being hosted.

The first thing we need for good face swapping is of course a good collection of faces. However, unlike our previous experience in using the CelebA dataset, we want specific faces of two individuals. One individual set of faces will be representative of our target face A, and the second face will be the face we want to swap with, B.

Our other requirement for gathering these faces is that we want them to be varied with different poses and lighting conditions. To keep things simple, we will also want to avoid faces with too much variation in eyewear, facial hair, and makeup. We typically don't need to worry about hair since our focus in swapping is often just the face.

For the next exercise, we are going to look at a tool that can quickly provide us with a collection of faces from specific celebrities. This tool is called Bing Image Downloader, and it is a Python package that allows you to do image searches and retrieve the results. Open your browser and let's jump into Exercise 9-1.

EXERCISE 9-1. DOWNLOADING CELEBRITY FACES

1. Open the GEN_9_Faces.ipynb notebook from the GitHub project site. If you are unsure how, then consult Appendix B.

2. The first cell in the notebook installs the package and requirements with the following code:

```
%%bash
git clone https://github.com/gurugaurav/bing_image_downloader
cd bing_image_downloader
pip install .
```

3. The %%bash at the top of the cell allows all the following commands to be executed directly against the Bash shell of the server runtime.

4. After that, we have a block that defines a form for easy selection of some celebrities and the number of faces to extract:

```
#@title DOWNLOAD IMAGES { run: "auto" }
search = "julia roberts" #@param ["will ferrel", "ben stiller", "owen
wilson", "eugene levy glasses","julia roberts","scarlett johansson"]
image_cnt = 1001 #@param {type:"slider", min:1, max:10000, step:100}
```

5. The `search` variable defines the search string we will use to find the faces in Bing. `image_cnt` is a value for the number of faces to download controlled by a slider. You typically want 500 to 1,000 good faces to train your face swap model on later.

6. Just below that, the next section of code does all the work of downloading the images to the given folder:

```
from pathlib import Path
from bing_image_downloader import downloader

folder = Path("dataset")
downloader.download(search, limit=image_cnt, output_dir=folder, adult_
filter_off=False, force_replace=True)
output_folder = folder / search.replace(" ", "\ ")
filter = output_folder / "*.jpg"
zip_file = search.replace(" ", "_") + ".zip"
print(filter,zip_file)
```

7. The previous code may take some time to download 1,000 images. Just be patient and wait until all the files are downloaded into the designated folder.

8. After that, the files are zipped up using the Bash `zip` command. Notice the use of $ before the predefined variables, allowing us to substitute Python variables into shell commands.

```
!zip $zip_file $filter
```

9. Then we can download the ZIP file using the code in the last cell. Running this cell will download the ZIP file to your machine.

```
from google.colab import files
files.download(zip_file)
```

You should run the previous exercise for the celebrities suggested on the form, or feel free to add your own. The faces/people we choose here will be dependent on the video we decide to deepfake later. If you are unsure of what faces to download, jump to the next section to download the subject video, which may help you decide which celebrities to search for.

Downloading YouTube Videos for Deepfakes

You may decide to just perform face swaps on single images, but in most cases, you will likely want to create deepfakes with videos. There are plenty of sources for free and paid videos you can access, but the easiest and best resource is of course YouTube. By downloading videos from YouTube, you often also have choices on the format of the content.

When you decide on the type of video you want as your subject for a deepfake, you often want to consider a few details, as outlined in the list here:

- **Short**: Choose videos typically less than 30 seconds in length. If you do want to deepfake larger videos, then you are likely better off using a different source than YouTube.

- **Easily recognizable faces**: Make sure to select videos with prominent and recognizable faces. It will be better if the individual's face is the focus for several frames as well.

- **Fewer faces**: Avoid videos with crowded scenes at least until you get good at perfecting the FS/deepfake workflow. More faces in a frame of video represents more work for you later. Faceswap has some useful tools to help manage this, but it generally works better to use fewer faces.

- **Popularity**: You will often want to choose the subjects of your video by popularity. The reason for this is the need for those initial 1,000 or so faces you will use for training. Training the Faceswap model later will depend on the quality of those faces you extracted in the previous exercise. Therefore, it is helpful to have popular enough celebrities to get a good set of diverse training images.

As you get more experienced with FS, you may want to break some or all those rules. You may just want to use your own home videos swapped with a celebrity for your personal enjoyment. At this point, those details are all up to you, and the methods presented here are only to be used as a guide.

For the next step in our process, we are going to use another third-party tool, YouTube-Downloader, to automate the downloading and packaging of videos from YouTube. Exercise 9-2 provides several sample video options, but you can easily swap out your own choices later. Open your browser again and jump on Colab to download a selection of subject videos.

EXERCISE 9-2. DOWNLOADING YOUTUBE SUBJECT VIDEOS

1. Open the GEN_9_Video_DL.ipynb notebook from the GitHub project site. If you are unsure how, then consult Appendix B.

2. The first cell in the notebook installs the youtube-dl package and requirements with the following code:

   ```
   !pip install --upgrade youtube-dl
   ```

3. After installing the package, we set up a form to select from a choice of subject videos and download them with the following:

   ```
   #@title SELECT VIDEO
   video = "Anchorman" #@param ["Elf", "Zoolander", "Schitts
   Creek","Pretty Woman","Avengers","Anchorman"]
   from __future__ import unicode_literals
   import youtube_dl

   videos = { "Elf" : { "url" : 'https://www.youtube.com/
   watch?v=3Eto6DU_2oI'},
               "Zoolander" : { "url" : "https://www.youtube.com/
               watch?v=KeX9BXnD6D4"},
               "Schitts Creek" : { "url" : "https://www.youtube.com/
               watch?v=hg1Uk60rBsc"},
               "Pretty Woman" : { "url" : "https://www.youtube.com/
               watch?v=1_TZEsUhXRs"},
               "Avengers" : { "url" : "https://www.youtube.com/
               watch?v=JyyGJk51n-0"},
   ```

```
        "Anchorman" : { "url" : "https://www.youtube.com/
        watch?v=88zGzznpnis"},
        }
video_url = videos[video]['url']

download_options = {}
download = youtube_dl.YoutubeDL(download_options)
info_dict = download.extract_info(video_url, download=False)
formats = info_dict.get('formats',None)

for f in formats:
  if f.get('format_note',None) == '480p':
    url = f.get('url',None)

print(url)
```

4. When this cell is done running, confirm that the URL is printed to the output
    window below the cell. If no URL is printed, then the video format is not
    supported at 480p, our preferred format. An alternative is to select another
    video or modify the format.

5. Next, we will do some imports for OpenCV2, an image and video library, to turn
    the download back into a video using our preferred codec. This step sets up the
    capture for video and does an initial frame count.

```
import cv2

input_movie = cv2.VideoCapture(url)
length = int(input_movie.get(cv2.CAP_PROP_FRAME_COUNT))
print(length)
```

6. From here, we want to render the captured video frames from YouTube
    back into a video using a codec of our choosing. Codecs are a form of video
    encrypting, and there are numerous options to choose from. You can either
    use the default codec selected or comment out the line and select another as
    appropriate. If after you download the video it is unable to play on your desktop,
    you may want to swap the codec and render the video again:

```
from google.colab.patches import cv2_imshow
from IPython.display import clear_output
import time
```

```
frame_width = int(input_movie.get(cv2.CAP_PROP_FRAME_WIDTH))
frame_height = int(input_movie.get(cv2.CAP_PROP_FRAME_HEIGHT))
print(frame_width, frame_height)

frame_number = 0
frame_limit = 1000
# Define the codec and create VideoWriter object
#fourcc = cv2.VideoWriter_fourcc(*'FFV1')
fourcc = cv2.VideoWriter_fourcc(*'XVID')
#fourcc = cv2.VideoWriter_fourcc(*'DIVX')
#fourcc = cv2.VideoWriter_fourcc(*'DIV3')
#fourcc = cv2.VideoWriter_fourcc('F','M','P','4')
#fourcc = cv2.VideoWriter_fourcc('D','I','V','X')
#fourcc = cv2.VideoWriter_fourcc('D','I','V','3')
#fourcc = cv2.VideoWriter_fourcc('F','F','V','1')

filename = f"{video}.avi"
out = cv2.VideoWriter(filename,fourcc, 20.0, (frame_width,frame_height))
while True:
    ret, frame = input_movie.read()
    frame_number += 1

    if not ret or frame_number > frame_limit:
        break
    out.write(frame)
    if frame_number < 10:
        cv2_imshow(frame)
input_movie.release()
out.release()
cv2.destroyAllWindows()
```

7. The cv2.VideoWriter constructs the writer to perform the rendering of the captured frames. We limit the number of frames rendered by the variable frame_limit, which is currently set to 1000. You can of course alter this as you see fit. As this writer is processing, it will also output the first 10 frames of video for inspection.

8. We next download the rendered video using the same block of code in the previous exercise:

```
from google.colab import files
files.download(filename)
```

9. When all the cells are run in the notebook, you should have a video downloaded to your machine in your preferred download folder. Be sure to watch the video to confirm it is the correct format and covers the right subjects. Note, we have plenty of opportunity to trim out areas of the video we may not want later. So, don't worry too much about excess content, but be sure your preferred subject is clearly visible and in frame for several seconds.

After you have the subject video and two sets of celebrity faces downloaded as zip files into a folder, unpack the files into new folders. Be sure that only a single set of celebrity images is in each folder and the video is also in separate folder. We will look at other organizational tips when we move on to the FS workflow in the next section.

Understanding the Deepfakes Workflow

Constructing deepfake videos or other content is relatively simple with a good tool like Faceswap or DFL. There is often a well-defined workflow that needs to be followed to prepare, mark, and align the content used in FS. For Faceswap, the basic workflow is shown here:

1. **Extracting**: The downloaded celebrity images and video need to be processed for facial extraction and alignment. This requires running the Faceswap software over either a folder of images or videos to first extract each face from an image. When a face is extracted, it is oriented correctly and then dropped into a new folder, with the description of the face added to an alignments file.

2. **Sorting**: After extracting faces, we move on to sorting those faces into a new folder. Faceswap has a tool to sort faces according to similarity and results in a neatly categorized folder of faces. By sorting the images, we can prepare for the next step trimming.

3. **Trimming**: After faces are sorted, we can review the images and remove any that don't match our primary subject. The software is not smart enough to extract the correct faces, and it will generally extract all faces in images or video frames. As such, we want to go though and remove those unneeded or unwanted faces.

4. **Realigning**: The alignments file generated from the initial extraction is critical to training and conversion later. After we remove the unwanted faces in the trimming step, we need to clean up the alignments file of any unneeded references. This step can be quickly performed with a tool within the software, as we will see later.

5. **Repeat**: You need to repeat the previous four steps for each of the celebrity subjects A and B as well as the desired output video.

6. **Training**: After the faces for the two subject celebrities are extracted, sorted, and trimmed, we can move on to training the A/B faces. Faceswap makes this especially easy provided you select the correct options.

7. **Converting**: When everything is cleaned and we have a well-trained model, we can move on to converting the video into a deepfake. This step is quick to perform provided the video length is small.

8. **Rebuilding**: Finally, when the conversion process is complete, we need to convert the deepfake images back to a video. The software has a tool to perform that as well, as we have provided another Colab notebook tool to do so.

Before we get into running the workflow with Faceswap, you will need to download the GUI client from the GitHub site at `https://github.com/deepfakes/faceswap/releases` where a Windows and Linux version is freely available. Be sure to download the version for your OS and then run it through the installer.

Installing Faceswap on Windows/Linux should be straightforward, and it includes all the required dependencies as well as setting up any needed GPU Cuda support for your graphics card. If you have ever tried to install a Python framework like PyTorch or TensorFlow on Windows, you know how difficult and convoluted the process can be. Fortunately, the Faceswap installer handles this all for us.

After you have the software installed, we can move on to our first step in the process in the next section. We will treat each step in the FS and deepfakes process as an exercise.

Extracting Faces

The process of extracting and identifying faces is well-defined and has been used for years. In this step, our goal is to extract each face in the various images and frames into separate files with just an aligned face. In the process, each face is entered into an alignments file describing the orientation and feature points.

Figure 9-3 shows the 68 feature points extracted from a face and entered into the alignments file. This file is critical to every step in the process as it defines the important features of each face. The actual alignments file is binary and not directly editable without tools like Faceswap.

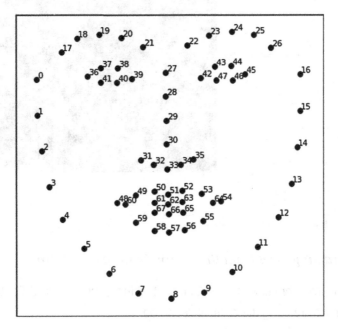

Figure 9-3. *68 facial feature points as described in an alignments file*

In Exercise 9-3, we start by extracting the faces from one of our earlier downloaded videos, but the same process will apply to a folder of images for the celebrity subjects. The output of this process will be the folder of extracted faces and the alignments file. Start up the Faceswap software and begin the exercise.

EXERCISE 9-3. EXTRACTING FACES

1. After the software starts, you want to make sure you are looking at the Extract tab, as shown in Figure 9-4.

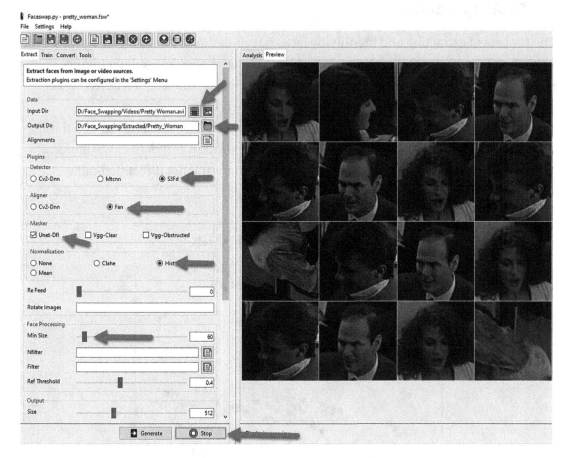

Figure 9-4. *Extracting faces from the Pretty Woman video clip*

2. Select the folder or file with your input content in the Input Dir field. For this example, we are using the downloaded video.

3. Choose an empty output folder where the extracted faces will be placed and put it in the Output Dir field.

4. Choose the Detector type to use for identifying the images. S3Fd is currently the best for extraction and the one we will use. If you want to learn more about the various options, check out the Faceswap forums at https://forum. faceswap.dev/.

5. The next option, Aligner, allows you to choose which plugin for orienting the images. Again, we will use the current best, Fan.

6. Masking is next and is the process of removing unwanted content from around the face and making sure features of the face are visible. For the masker, we will use the Unet-Dfl model.

7. As always, we then want to normalize the extracted images for better training later. Hist is the suggested normalization method and appears to work well for training all the models.

8. The last option we want to set is Face Processing. This set of options is advanced aside from the minimum size of the faces we want to extract. Depending on your content, you may want to adjust this number to a nonzero value that represents the minimum face size you want to extract. You may want to play with this value for various content depending on your final production needs.

9. Finally, we can start converting by clicking the Convert button at the bottom of the window. After clicking Convert, you will see log output in the bottom window, and then on your right you will see the various faces as they are being extracted. When the process is complete, you should have a folder filled with faces and a new alignments file in your original input folder.

You will need to run the extraction process for the subject video and the two celebrity sources you want to swap. After you complete the extraction for all content, we can move on to the next section about sorting.

Sorting and Trimming Faces

Not all the faces we extracted in the previous step will be useful and may even be of the wrong subject. This will be especially true if you have converted a video clip with several actors in a scene. What we want to do, therefore, is remove unwanted faces from the extracted folder and then rebuild the alignments file.

To do that, we need to manually go in and remove unwanted images from the extracted folder. Since many of these images will often be mixed, a useful process is to first sort those images by similarity using a tool built into the software.

In Exercise 9-4, we will run the sorting tool to sort the images by similarity and then remove those unwanted faces from the folder. This is a fairly simple process you will need to perform for each extracted face set. Let's begin by sorting the images from an extract folder next.

EXERCISE 9-4. SORTING AND REMOVING FACES

1. With the software running, open the Tools tab and then the Sort tab, as shown in Figure 9-5.

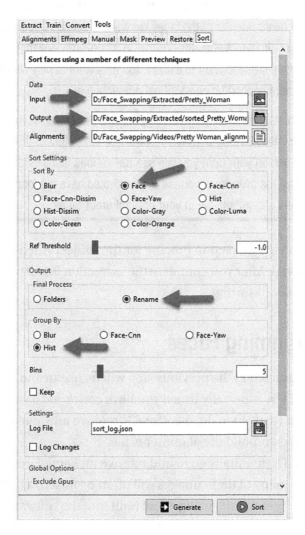

Figure 9-5. *Using the Tools/Sort feature to order faces*

2. Under the Data group, we need to again fill in the Input, Output, and Alignments file locations. The input folder is the location of the extracted images. The output folder is a new folder you want to contain the sorted images, and the alignments file will be found in your original input folder's source.

3. Within the Sort Settings and Sort By group boxes, you want to select Face as the method of sorting.

4. For the Output/Final Process group, you want to keep the default using the Rename option. This will rename the images based on the sort order.

5. Then under Group By, we will use Hist as the default but preferred method.

6. Finally, click the Sort button at the bottom of the window, and in a few minutes, you will be notified that the output folder has the sorted faces.

7. Locate and open the output folder, as shown in Figure 9-6, where the faces are sorted by similarity.

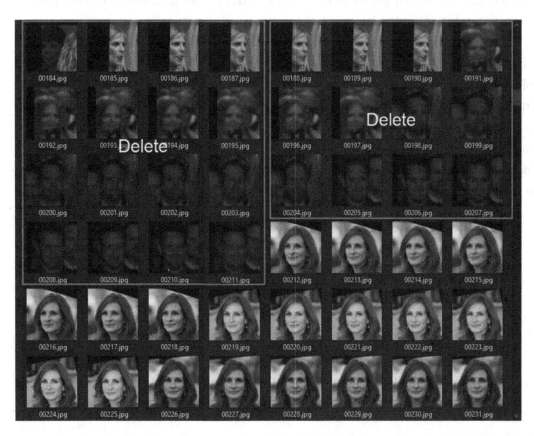

Figure 9-6. *Example of sorted output from the Julia Roberts extracted celebrity images*

8. At one end of the folder, you will see faces and garbage images you likely don't want to train on or swap. Go ahead and delete those images. You may also want to remove faces that are abnormal. Perhaps the celebrity is wearing sunglasses or had unusual facial hair or makeup. You may notice several duplicate images, which is okay but not ideal.

With the unwanted faces removed, we now need to move and clean up the alignments file using another tool in the software, as we will see in the next section.

Realigning the Alignments File

The alignments file is used to identify the 68 mapping points of a face in an image and is heavily used to define which face is the target in the image. Often you may extract images of faces with more than one face, and the alignments file helps the software identify which to use. It also helps to mask the face for better mapping of one face to another when we apply conversion later.

After we have the unwanted faces removed from the folder, we need to rebuild the alignments file. Specifically, we need to remove the faces we deleted from the sorted folder from the alignments file. Otherwise, the software will complain we are missing images when we proceed to training.

The process is quite simple for removing faces from the alignments file and is included as a tool within the software on the Tools tab. You will need to clean up each alignments file that you sort and remove images from the extract folder. Return to the Faceswap software, and we will go through the process of removing faces in Exercise 9-5.

EXERCISE 9-5. REMOVING FACES FROM THE ALIGNMENTS FILE

1. With the software running, open the Tools tab and then the Alignments tab, as shown in Figure 9-7.

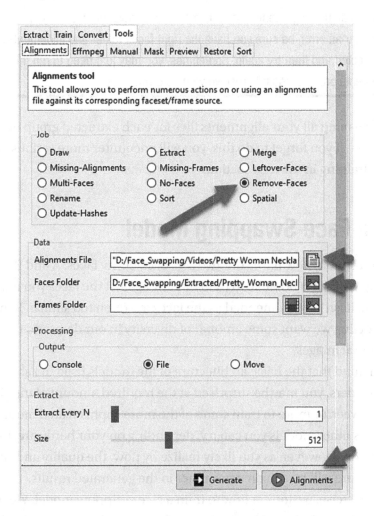

Figure 9-7. *Removing faces from an alignments file*

2. In the Job group box, select the Remove-Faces option. As you can see, there are several operations you can perform on an alignments file. For more information on how use these tools, consult the Faceswap repository or forums.

3. Next in the Data group box, select the source alignments file and the folder you extracted, sorted, and removed faces from in the Faces Folder field.

4. Leave the rest of the defaults as they are, but do make sure they are set the same way as in Figure 9-7.

5. Finally, click the Alignments button to start the removal process. If you encounter an error, be sure you have the right faces folder and alignments file selected and then try again. Sometimes, you may just need to run the alignments process again to resolve with no errors.

Be sure to clean up all your alignments files for each extracted group of images and video you process. If you forget to do this, you will encounter more serious errors when we move on to training in the next section.

Training a Face Swapping Model

At this point you should now have your two sets of celebrity faces, A and B, that we will use for training. While it is possible to train directly against the faces you may want to swap for a video, it isn't recommended as the faces are often too similar. Remember in generative modeling we want some amount of diversity in our data, and that means it's best to use external images.

Keeping in mind that the basic architecture of the models used in Faceswap are double autoencoders, you may be surprised at the required amount of training time. It can take up to a week or more to train some of the more sophisticated models like Villan. There are various other models you can try, depending on your hardware, including a lightweight version. However, as you likely realize by now, the quality and size of the model's architecture will be a main determiner in the generated results.

In the next exercise, we look at how to train a face swapping model. We have already done various similar trainings when we explored image to image paired and unpaired translations with GANs in Chapters 6 and 7. The Faceswap software makes the process of setting up training relatively easy, but it may take some time to master which model works well for your hardware and what results you ultimately want. We will continue where we left off from the previous exercise and start training a model in Exercise 9-6.

EXERCISE 9-6. TRAINING A FACE SWAPPING MODEL

1. With the software open, select the Train tab, as shown in Figure 9-8.

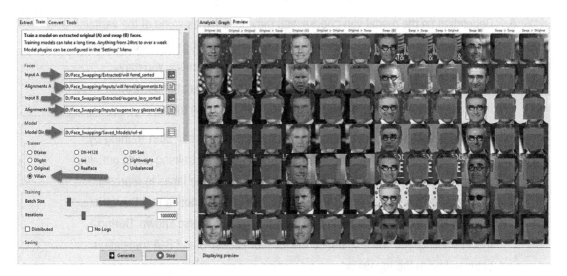

Figure 9-8. *Training a face swapping model*

2. The first thing we will do is set the Input A and Input B fields to the folders where we have the extracted, sorted, and trimmed faces. Figure 9-8 shows an example using Will Ferrell as Input A and Eugene Levy with glasses as Input B. Care was taken to remove images of Will Ferrell with a mustache made famous for his role in *Anchorman*. Faces of Eugene Levy without glasses were also removed. Deciding which set of faces to remove will depend on the source video you want to convert. If we want to convert images of Eugene Levy in *Schitt's Creek*, we would likely not use his famous glasses look.

3. Next, we need to set the Alignment A and Alignment B file sources. Again, these files will typically be in the original input folders.

4. Setting the model in the Model Dir field is next, and it is best to use a new folder for this. This will be the folder where the trained models will be saved as training progresses.

CHAPTER 9   DEEPFAKES AND FACE SWAPPING

5. From here, we need to determine which model to use from the Trainer group box. The Villan model selected is considered the best but also requires the most horsepower for training. If your computer is lacking a supported GPU, then it may be better to start with the Lightweight or Original model.

6. Depending on the model you choose, you may need tweak some of the various hyperparameters like "Batch size" found under the Training group box. When using the Villian model, which consumes a lot of GPU memory, you may need to reduce the batch size from the default of 16. You can modify other hyperparameters by accessing the settings from the main menu.

7. After you are done filling in the basic required fields, you can start training by clicking the Train button at the bottom of the window. As training starts out, you will see as shown in Figure 9-8 the use of masks over the faces to represent the areas the model is training to translate. You will also notice that the various model swaps from Input A to Input B, and the reverse is also shown. During model training, you will also notice the loss for each A-B and B-A translation being output in the log window.

8. As mentioned previously, it may take a substantial amount of time to train a model accurately. While the model is training, you can view the loss progress by clicking the Graph tab found beside the Preview tab in the training window.

9. You can at any point stop the training by clicking the Stop button and then move on to testing a conversion. Generally, you will want your model to display accurate translations, as shown in Figure 9-9, which shows a later state of training, before continuing to create a deepfake video.

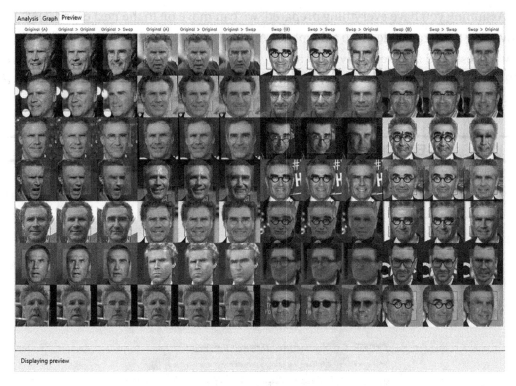

Figure 9-9. *Later stages of model training using Villan*

When you are happy with the trained model, you can move on to converting your subject video into a deepfake. Of course, you don't have to wait for the model to complete training, but realize partially trained models may generate poor results. The great thing about having a fully trained model for face swapping is that it can be used across multiple videos that feature the same subject. Keep in mind, though, that you need to account for actors appearing with makeup, facial hair, and glasses that are outside their normal appearance.

With a sufficiently trained model, we can move on to the next section and learn how to generate a full deepfake video.

Creating a Deepfake Video

Getting a believable and accurate deepfake video is mostly about the performance of the model you trained in your previous step. It is also important that you have extracted and cleaned the faces from the video you will be converting. Be sure before proceeding that you have extracted, sorted, and trimmed the faces in your target conversion video.

279

Assuming you have everything prepared, we can move on to the next exercise of converting the target video into a deepfake. Again, Faceswap makes this an extremely easy process, and a 30-second video can be converted in less than 5 minutes if you have a GPU-supported environment. Jump back into the software, and we can work toward to creating your first deepfake video; see Exercise 9-7.

EXERCISE 9-7. CONVERTING A VIDEO TO SWAP FACES

1.   With the software open, select the Convert tab, as shown in Figure 9-10.

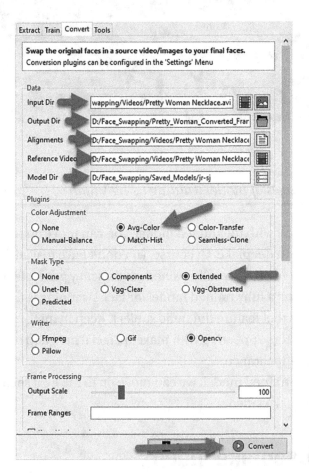

Figure 9-10. *Interface for converting a video to deepfake frames*

2. There are several options available to modify the results of a conversion. We will stick to the basics in this exercise, but you can learn more about the various options from the Faceswap repository or forum.

3. As typical, we start with the Data group box at the top of the window. You will need to set the Input Dir option to the actual target video you want to convert. Then set the Output Dir option to a new folder where the video frames will be written to. Faceswap does not create the output video by default, and combining the frames will be done in the last phase.

4. Next, we can set the alignments file in the Alignments field. In addition, the Reference Video input will the same as the Input Dir video. Then finish off this first section of inputs by selecting the model folder you used for training.

5. Typically, you can leave the Color Adjustment section as the default. However, if you notice differences in color around the face of a converted image, you may want to play with these options further.

6. The Mask Type setting defines the mask used for the area around the face that needs to be converted. Different masks will alter that area using several factors dependent on the model plugin. If you notice the face not getting completely translated in your converted output, you likely need to adjust the mask used in the conversion process. You may also need to select and define different masks used for the extraction process. These are advanced settings and again best documented on the Faceswap site.

7. You can stick with the other default settings, as shown in Figure 9-10. When you are ready, you can start the conversion process by clicking the Convert button at the bottom of the window.

After the video has been converted, you can open the Output Dir folder and review how well the process worked. If you find you are not happy with the conversion, start by going back and training the model more or with a different type. You may also need to understand some of the finer options around masking and color normalization. Again, consult the Faceswap documentation, forum site, and other sources for help.

When you are happy with the conversion, we can move on to the last step of creating the final deepfake video from the converted frames in the next section.

Encoding the Video

The last step in creating a deepfake is taking all the converted video frames and combining them into a video. Faceswap has a tool that can convert frames to video, but it relies on the reference video for extracting the codec and other properties. However, many times when creating a deepfake you may want to use only a partial set of frames to rebuild the video.

Fortunately, Python and OpenCV make constructing a new video with just fames quite easy as we already seen with the YouTube-Downloader exercise. Therefore, it is a simple matter to set up and use another Colab notebook that can easily do the video creation for us using the frames and codec of our choosing.

In the last exercise of this chapter, we jump back on to Colab to use a Python tool to create the deepfake video. If you don't want to use a notebook and you have a proper Python environment set up, you can also copy the code to a local file and run it directly from your machine. Either way, the flexibility of using this notebook should make it easy to quickly build a deepfake; see Exercise 9-8.

EXERCISE 9-8. CREATING THE DEEPFAKE VIDEO

---

1. Open the GEN_9_Make_Video.ipynb notebook from the GitHub project site. If you are unsure how, then consult Appendix B.

2. Open the folder/files tab on the left side of the window and create a new folder called images. Using the ellipsis menu beside the new folder, click the Upload option to open a file browser view.

3. Locate and open the Output Dir folder of the converted frames from the previous exercise. Select some of the frames you want to convert to a video using the File Browser. When all the images are selected, click Open in the browser to upload the images. Wait for all the images to uploaded before continuing.

4. Return to the notebook cells, and you can now run the entire notebook from the menu by selecting Runtime ➤ Run all.

5. The first block of code, shown here, pulls all the images from the folder and puts them in an array/list for later processing:

```
import cv2
import numpy as np
import glob

img_array = []
for filename in sorted(glob.glob('/content/images/*.png')):
    print(filename)
    img = cv2.imread(filename)
    height, width, layers = img.shape
    size = (width,height)
    img_array.append(img)
```

6. Notice the use of the glob and sorted functions to load all the images and sort them.

7. Moving down to the next cell, we can see how easy it is for Python with OpenCV to create a new video file. Again, if you need to modify the codec used to create the video, uncomment/comment the appropriate lines, as shown in the following code:

```
movie = 'deepfakes.mp4'
#codecs
# Define the codec and create VideoWriter object
#fourcc = cv2.VideoWriter_fourcc(*'FFV1')
#fourcc = cv2.VideoWriter_fourcc(*'XVID')
#fourcc = cv2.VideoWriter_fourcc(*'DIVX')
#fourcc = cv2.VideoWriter_fourcc(*'DIV3')
fourcc = cv2.VideoWriter_fourcc('F','M','P','4')
#fourcc = cv2.VideoWriter_fourcc('D','I','V','X')
#fourcc = cv2.VideoWriter_fourcc('D','I','V','3')
#fourcc = cv2.VideoWriter_fourcc('F','F','V','1')
out = cv2.VideoWriter(movie,fourcc, 15, size)

for i in range(len(img_array)):
    out.write(img_array[i])
out.release()
```

8. In the video creation code, you can also see where the `cv2.VideoWriter` function is called to define the movie name, type of codec, frames per second playback, and video size.

9. Finally, in the last block of code, the video will be downloaded using the same pattern as we did earlier in this chapter. After the movie has been built and downloaded, review it and decide if you are happy or need to go back to the drawing board.

With your video encoded and now playable, assuming you used the right codec, you can enjoy the satisfaction of seeing your first complete deepfake. You may find that after your first attempt, the results are less than optimal. As mentioned previously, there are plenty of modifications you can make through the FS workflow, training, and conversion process.

At this point, you can move on to creating various other deepfakes or improve on ones you already attempted. You could also try different or more complex variations. Perhaps attempt to do multiple subjects in a video, which would require you to run multiple passes of conversion across the same output video. There are more advanced workflows as well that will allow you to do multiple subjects, but we leave that up to you to explore on your own.

You could also move on to exploring the other software options for face swapping and creating deepfakes we covered early on in this chapter. Just be aware that no matter which software you are working with, you still want to maintain those ethical standards when creating fake videos. Either way, there are plenty of options for you to explore on your own.

Conclusion

Probably the most controversial and abused form of generative modeling is the application of face swapping and the creation of deepfakes. In many aspects, the ethical abuse of this subset of generators often puts a black mark against the whole technology. There is also potential for the abuse of this technology for creating disinformation and fake news. It is for those reasons that it is important you follow the ethical guidelines when performing face swapping.

While it remains to be seen how this technology will evolve, there is already a certain amount of fear of misuse for those who practice this craft. It certainly is likely that our ability to create better face swaps will improve. After all, the models used for current face swapping workflows are far less advanced than many of the generative models we used in this book.

However, there also remains several opportunities to extend this technology outside of just faces. Imagine swapping workflows that can alter people's clothes, the background, and even the style of a movie. Perhaps one day we will see movies like *Star Wars* converted to westerns or perhaps famous westerns translated to space operas.

Face swapping and deepfakes are likely here to stay if most of us develop them in an ethical manner. If this craft continues to be abused for the creation of fake porn and fake news, we could risk government controls and legislation against such tools. Perhaps even worse would be the tarnished reputation of generative modeling itself.

In the next and final chapter of this book, we step back and explore how we can identify content that has been created from generative modeling. There is a major focus around this area due to security concerns regarding deepfakes and the potential abuse for creating fake news. As such, understanding how we can identify generated fakes will be good final step to complete our journey.

## CHAPTER 10

Cracking Deepfakes

Throughout this book we have considered numerous strategies for generating content that will match the reality we know or perhaps even surpass it. While most of that time has been spent exploring the realm of faces and creating realistic faces, the same techniques can be applied to any other domain as we see fit. However, it is perhaps being able to generate realistic faces that is the most frightening to so many.

Faces are, after all, the heart and soul of how we communicate and convey emotions. Being able to create real faces or swap faces, as it were, opens a world of possibilities for abuse in all manner of things. As we talked about in Chapter 9, this is an ever-present danger and one we need to take seriously in terms of ethics and understanding.

In the Canadian Navy, all vessels are equipped with a ship's diver, and these individuals are trained in all manner of demolitions from construction to deactivation. The course is so extensive it is said that if you pass, you become a demolitions expert, and if you fail, a terrorist. While certainly this isn't true, it is said because individuals who fail the course learn how to build a bomb but likely failed how to disarm it.

We could perhaps apply the same analogy to a generative modeler who builds deepfakes and face swapping but fails to understand the principles of detecting them. As we saw in the previous chapter, it is relatively easy to create a deepfake, and you don't really need to understand the finer details of generators. Now you likely won't be classified as a terrorist if you fail generative modeling, so this may not concern you.

However, if you do understand generators and how they fail, or better yet succeed, you can also find their faults. After all, for most of this book, we looked at better ways to fool a discriminator so well that it could fool ourselves. Now it is time to flip the script and look at ways at which we can detect the fakes using the knowledge we have gained.

Understanding fake content detection will provide us with not only a new set of tools but also opportunities. It is likely that the future will be littered with all manner of fake content that will need to be policed at all levels. From governments to private industry, there likely will evolve a whole new machine learning developer/engineering position.

© Micheal Lanham 2021
M. Lanham, *Generating a New Reality*, https://doi.org/10.1007/978-1-4842-7092-9_10

In this chapter, we look at how to detect fake content and specifically deepfakes. We will first look at the various methods used to manipulate content/faces with generative modeling. From there we will look at the strategies used to determine if content is fake and how they can be applied to determining what is real. We finish the chapter by looking at several useful tools and methods for identifying the current fakes.

While we won't have the time to explore the various strategies and techniques for detecting fakes and deepfakes as we have previously, this chapter should provide the interested reader with a good foundation for how to pursue this subject further. Here are the highlights we will cover in this chapter:

- Understanding face manipulation methods

- Techniques for cracking fakes

- Identifying fakes in deepfakes

The field of fakes and deepfakes detection is just emerging and will likely significantly mature in the next few years. This chapter is meant as a high-level introduction to a quickly expanding field that will evolve into advanced methods and tools. In the next section, we look at the methods and strategies we use detect fakes, which will be the basis for those tools.

Understanding Face Manipulation Methods

In the realm of facial manipulation, we typically categorize the techniques into two broad categories. There is identity to identity transformation where one individual's face is swapped for another. The other form is where an identity is translated to the pose of another identity, called *facial reenactment*.

We already looked at the identity to identity transformation method called face swapping in Chapter 9 and used that to construct a deepfake. While we used deep learning dual autoencoders to project the facial transformation, there are in fact other graphics-based methods. Here is a summary of the high-level identity transformation methods:

- **FaceSwap**: This is not to be confused with the software package Faceswap. FaceSwap is a graphics-based approach that is akin to pulling the face from the target image, projecting it to a 3D model using facial landmarks, and then capturing the replacement face. This method is currently used in the film industry for all manner of swapping faces and other manipulations.

- **Deepfakes**: This works by first extracting the faces from the source and target identities. A deep learning network is trained to transform face A into face B, and vice versa. The trained model can then be used to swap face A with face B using the network to interpret any required pose changes. Again, we covered this whole process in fine detail in Chapter 9.

Facial reenactment is the second category of facial manipulation and works by keeping the target identity but replacing the pose from another source, where the other source may be a person speaking. This method is often lumped in with deepfakes, but for our purposes we will define this as *facial reenactment* rather than face swapping (FS).

Figure 10-1 shows the difference between FS and facial reenactment (FE) or pose swapping using some well-known popular examples. The first top half of the figure demonstrates face swapping with the face of Lady Gaga and the actor Steve Buscemi. This swap was performed with the Villain model using Faceswap.

Figure 10-1. *Difference between face swap and pose swap*

In the bottom half of Figure 10-1, President Barack Obama's likeness is being used, but his movements and actions (pose) are being replaced by comedian Jordan Peele. The image is extracted from an entire video that demonstrates the full power of pose swapping. During the video, the comedian goes on to speak as the president making various false claims.

Pose swapping is currently implemented using two diverse strategies, not unlike what we see with FS. There is a graphics-based approach and a deep learning method that can be applied to swap the pose of a target. Here are more details about each method:

- **Face2Face**: This is a graphics-based method like FaceSwap but uses two input video streams where key frames are extracted and used as blend targets. Instead of extracting the face from landmarks, this time the pose is extracted. Then the pose is translated to a 3D feature map used to blend the target face.

- **NeuralTextures**: This method uses a typical GAN not unlike Pix2Pix or another image to image translation model to translate the pose from one image to another. There are plenty examples of various open source projects that demonstrate this technique, but the current top contender is Avatarify. With Avatarify, using the right hardware, you can translate your pose in real time or the faces of Mona Lisa or Albert Einstein, among others. If you need to remember how this method works, look back to Chapters 6 and 7 where we trained models on paired and unpaired images for image to image translation.

Figure 10-2 was extracted from the paper "FaceForensics++: Learning to Detect Manipulated Facial Images," which is a great source of work showcasing various facial detection techniques. The authors have even been so gracious as to provide a free source of facial manipulated images and videos on their GitHub repository at `https://github.com/ondyari/FaceForensics/`.

Figure 10-2. *FaceForensics++ breakdown comparison of facial manipulation methods*

It should be noted that the authors in the FaceForensics++ paper make the distinction to call face swapping *deepfakes*, as shown in Figure 10-2. We will also now make that distinction and from now on call face swapping fakes *deepfakes*, leaving the broader deepfakes definition to represent the genre of facial manipulation content.

FURTHER REFERENCES

Exploiting Visual Artifacts to Expose Deepfakes and Face Manipulations:

https://ieeexplore.ieee.org/abstract/document/8638330

DeepFakes: a New Threat to Face Recognition? Assessment and Detection:

https://arxiv.org/abs/1812.08685

What do deepfakes mean for brands?

https://www.thedrum.com/opinion/2020/01/10/what-do-deepfakes-mean-brands

---

Now that we have clearer understanding of the scope of facial manipulation techniques, we can move on to understanding how to identify such fakes. In the next section, we introduce the various methods currently employed to identify deepfakes.

Techniques for Cracking Fakes

The research into developing methods to crack and identify deepfakes has flourished in recent years, certainly encouraged by the fear of increasingly better generative models. As we have seen throughout this book, the early days of generators could be easily identified and were of no concern. That has all changed in recent years, and now it is almost impossible for a human to distinguish what is a fake anymore with a mere glance.

However, and perhaps more fortunately for us and generative modeling in general, the same AI can be used against itself. After all, as we have seen throughout this book, we often struggle with the balance between the training of a discriminator and a generator; we may often have to handicap the discriminator so that a generator can produce great results.

In terms of cracking fakes, we then want to construct the best discriminator that can identify fakes using various feature detection methods. Of course, that doesn't mean that an excellent critic could not be used to adversarial train a better generator. In fact, there is plenty of research doing just that. How this arms race unfolds remains to be seen, but for now let's consider the methods used to identify fakes.

We can break down the current methods used to identify fakes into three categories, as defined in "DeepFakes and Beyond: A Survey of Face Manipulation and Fake Detection" by Ruben Tolosana et al. This excellent paper walks through the current categories of methods used to identify fakes and is summarized here:

- Handcrafted features

- Learning-based features

- Artifacts

The important thing to remember about each of these categories is they define broad techniques that may use a combination of AI/ML and feature extraction techniques through graphics packages like OpenCV. We will review each of these categories in more detail by starting in the next section.

Handcrafted Features

In many deepfake videos or images, there are often noticeable flaws in the reconstruction of features. If these flaws are especially evident, a human author may often edit them out with graphics software. Often, though, these flaws may not be so evident and are left in as markers of the fake.

Discovering these markers, our misaligned and constructed features can be accomplished with a variety of techniques that look to classify known features by similarity. A person, for instance, may tilt their head a certain way when talking or only open their mouth a specific distance. These features are specific to individuals and are difficult to mimic or translate effectively.

We can discover these hidden features of the way someone talks by handcrafting features using software like OpenCV to measure the amount of head tilt, eye movement, lip movement, and so on. Then we can compare like measurements using similarity feature maps to identify faces or facial posing that is consistent with an individual's identity.

Figure 10-3 was extracted from the paper "Protecting World Leaders Against Deep Fakes" by Shruti Agarwal and Hany Farid, in which the authors analyzed the use of handcrafted features to identify deepfakes. The figure shows a comparison of various famous speakers using the encoding of several handcrafted features. From the figure it becomes obvious to distinguish between the real Barack Obama and the lip-synced Jordan Peele version.

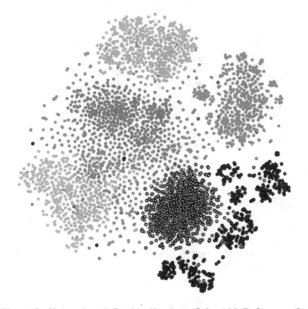

Figure 3. Shown is a 2-D visualization of the 190-D features for Hillary Clinton (brown), Barack Obama (light gray with a black border), Bernie Sanders (green), Donald Trump (orange), Elizabeth Warren (blue), random people [23] (pink), and lip-sync deep fake of Barack Obama (dark gray with a black border).

Figure 10-3. *Showing the use of handcrafted features to identify a speaker's identity*

While using handcrafted features is successful in demonstrating the obvious differences between various speakers, it still implements human bias. Ideally, what we want to do is remove human bias and allow a model to learn which features to compare on their own. This is the method we will look at in the next section.

Learning-Based Features

Just like the critics we develop with the adversarial training of GANs, we can take this concept further using convolution neural networks to identify and learn features that mark fakes. As showcased in the FaceForensics++ paper, there are several architectures that have been used to learn fake feature markers.

Figure 10-4 is another extracted image from the FaceForensics++ paper that demonstrates the differences between the various architectures used to identify fakes, including the use of handcrafted features (Steg. Features + SWM). The accuracy was measured on the raw images as well as high- and low-quality versions. The clear winner in all cases is the XceptionNet architecture, which we will review later in this chapter.

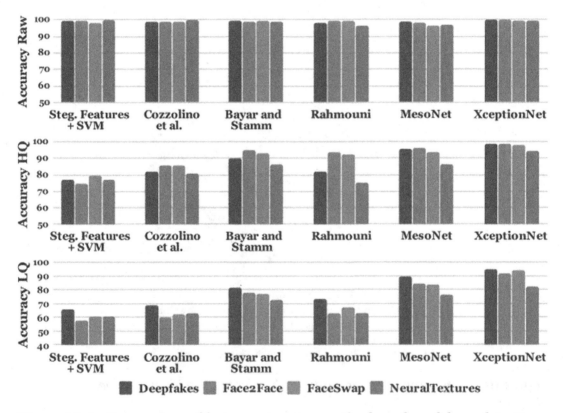

Figure 10-4. *Comparison of feature extraction methods and models on detection of fakes*

The construction of feature learning models using CNN is generally like that of the discriminator in a GAN. As we have seen, convolutional layers are the primary method of feature extraction found in most adversarial/GAN models, where the output of the model will typically be a binary classification identifying if the image is fake or real.

In the paper "GANprintR: Improved Fakes and Evaluation of the State of the Art in Face Manipulation Detection" by Joao C. Neves et al., the authors use a technique of reusing various CNN architectures like ExceptionNet as a learning decoder placed within an autoencoder.

Figure 10-5 is an image extracted from the GANprintR paper that describes the architecture of the autoencoder used to remove possible features that may identify the image as a fake. The model is trained by feeding real faces into the model and then decoding it with an XceptionNet or other CNN model that identifies fakes.

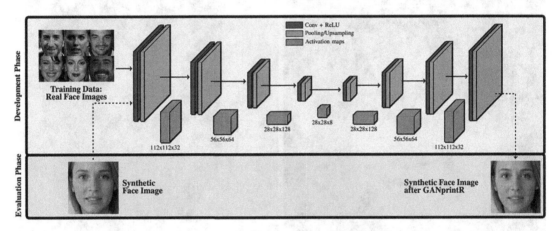

Figure 10-5. *Autoencoder designed to reverse feature anomalies in fakes*

The result of training real images to identify and remove fake markers again improves on the quality of the fake image. Fake images and content can then be fed into the model to likewise remove fake markers from synthetic images. This is just another example of the current arms war being played out on the field of facial manipulation.

Of course, as we have seen throughout this book, there are often other features outside the face or primary target that can clearly define evidence of faking. In the next section, we look at how these artifacts can be used to identify synthetic images.

Artifacts

Aside from understanding how well specific features are being correctly faked or translated, there is always the possibility of other artifacts identifying a fake. Recall our previous look at the site `https://www.whichfaceisreal.com/` that demonstrated the comparison of real faces beside those generated with StyleGAN.

Figure 10-6 was extracted from the whichfaceisreal.com site as an example of how well StyleGAN can generate a synthetic face image. By looking at the faces themselves, it is difficult to determine which face is real or fake. Often the trick is to spot the artifacts or the lack of background details from image to image.

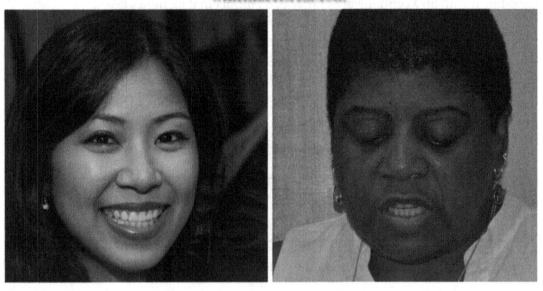

Figure 10-6. *Can you identify the real face?*

Developing methods that can identify these artifacts in images can also provide a basis for full or partial synthetic image identification. In the paper "Do GANs leave artificial fingerprints?" by Francesco Marra et al. (2019 IEEE Conference on Multimedia Information Processing and Retrieval [MIPR], IEEE Xplore, April 25, 2019, `https://ieeexplore.ieee.org/document/8695364`), the authors examined the artifacts left behind through the generation of synthetic images using CycleGAN, ProGAN, and others.

Figure 10-7 shows an example of the fingerprints the authors were able to identify for CycleGAN and the ProGAN from photo marks. Each photo or image can be analyzed using a photo-response nonuniformity (PRNU) pattern to determine if and how an image may have been manipulated.

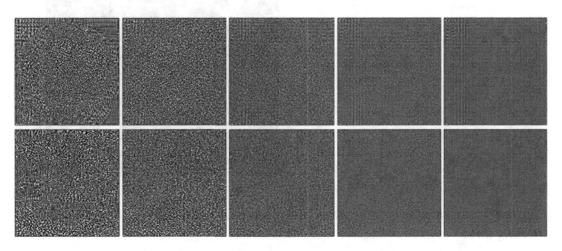

Cycle-GAN o2a (top) and Pro-GAN kitchen (bottom) fingerprints estimated with 2, 8, 32, 128, 512 residuals.

Figure 10-7. *Comparison of GAN fingerprints identified at various residuals*

Based on this study of the fingerprint of images in models, we can then understand which parts of image may have been manipulated. Looking at Figure 10-7, you can also visualize how the architecture of a GAN can contribute to this fingerprinting. Different architectures and feature extraction generation techniques will yield different fingerprints.

The authors went even further to compare various GAN architectures and real images taken with a physical camera. What they found, as shown in Figure 10-8, is that by comparing the correlations of fingerprints, there are easily distinguishable markers left behind depending on the type of GAN and the source set of images.

Architecture	Target / Camera model	Abbreviation
Cycle-GAN	apple2orange	C1
	horse2zebra	C2
	monet2photo	C3
	orange2apple	C4
	photo2Cezanne	C5
	photo2Monet	C6
	photo2Ukiyoe	C7
	photo2VanGogh	C8
	zebra2horse	C9
Pro-GAN	bedroom	G1
	bridge	G2
	church	G3
	kitchen	G4
	tower	G5
	celebA	G6
Star-GAN	black hair	S1
	blond hair	S2
	brown hair	S3
	male	S4
	smiling	S5
Camera	Nikon-D90	N1
	Nikon-D7000	N2

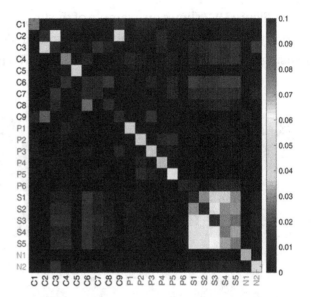

Figure 10-8. *Example of fingerprint correlations applied to GANs/camera and training data*

Note that the table and correlation plot in Figure 10-8 have an error where the corresponding ProGAN models need to be mapped from G1… to P1… in the figure. The comparison shows the average correlation between fingerprints, with the diagonal showing larger values for similar model comparisons between the same fingerprints.

Likewise, notice from Figure 10-8 how well the variations of the attributes a StarGAN is trained to replicate match to each other. We can also see how a CycleGAN is easily identifiable, but things get more difficult to discern with the ProGAN. Unfortunately, the others did not test the self-attention or the StyleGAN and StyleGAN2, which likely would have shown some remarkably interesting results.

In any case, what we can certainly learn is that there are identifiable artifacts left behind based on the type of GAN and the way it was trained. This is something you may have already realized, especially if you have gone through several of the exercises in this book. It can be obvious how the architecture of a GAN may leave indiscernible artifacts in the produced fake images.

While we leave it up to you to review all the papers mentioned in this chapter on your own, we will finish up the chapter by looking at some packages that may be used to identify various incarnations of fakes.

Identifying Fakes in Deepfakes

As we have seen throughout the previous section, there are a number of strategies and methods for identifying deepfakes and facial manipulation. We have looked at the broad categories of methods you can employ, but in this section we want to look at specific packages you can employ to identify reality.

To simplify things, we will look at the class of methods we identified earlier as feature learners—those that use CNN architecture and other layer types to learn the features that may identify synthetic images. The current best of these as identified in the FaceForensics++ paper is the XceptionNet, but we will look at others as well.

Here is a summary of a few better-known open source packages that may be used to identify deepfakes and other forms of synthetic images. These packages use a variety of methods from feature learners with PyTorch and feature extractors with OpenCV. All the code is in Python so you can review the source code on your own:

- **MesoNet-Pytorch** (https://github.com/HongguLiu/MesoNet-Pytorch): This is a CNN architecture-based model that learns features by being trained on fake images. Using this model requires a diverse set of fake images that you can pull from the FaceForensics++ site.

- **Deepfake-Detection** (https://github.com/HongguLiu/Deepfake-Detection): This is by the same team and author of the previous repository. This implementation extends the model and training architectures with new model variations and helpful ways to train and test models.

- **Pytorch-Xception** (https://github.com/hoya012/pytorch-Xception): This is the implementation of the Xception model using PyTorch in a Jupyter notebook. This example can be easily converted to Google Colab and trained and tested in the cloud using the provided example. The Xception model is the current gold standard in identifying deepfakes.

- **DeepFakeDetection** {https://github.com/cc-hpc-itwm/ DeepFakeDetection): This package uses the handcrafted feature extraction method with OpenCV to identify inconsistent features in generated images. The example in this project is developed on top of Jupyter Notebook, which should be easy to convert to Google Colab if you are so inclined. The added benefit of this approach is the lack of model training, which may be useful if you lack a good source of fake data.

These packages are all relatively easy to use and set up but, in some cases, require a substantial amount of training data. Again, a good source of this training data is the FaceForensics++ repository page, which has a good variety of fake synthetic content.

There is also a current trend in deepfakes detection where several variations of techniques and models may be combined to provide an overall fake score. That score can then represent a broader approximation of whether the faces or other content in images was faked. It also remains to be seen how such methods could be used in the reverse, that is, to generate better fake images.

Ultimately, the arms race between creating fakes and identifying them is only getting started, and who/what methods come out on top likely won't be decided for years. Unlike a military arms race, however, this race will certainly generate a plethora of new technologies and strategies that should allow generative modeling to become more mainstream and diverse. Ideally, this diversity can be applied across industries and applications.

Conclusion

The deepfakes arms race is only the start of an explosion in growth in the field of generative modeling. While we are still trying to understand the nuances of creating better synthetic content, there likely will always be the need to identify such fakes, whether the application is for the generation or manipulation of faces or other content.

What we found in this chapter is that the best method we have at our disposal currently is an implementation of an adversarial critic called XceptionNet. This model is built with a CNN architecture and resembles the discriminator in a GAN. However, instead of training such models with real and fake images, only fake images are used.

We also saw that as we increase our ability to identify fakes, we in turn provide new ways of critiquing the fakes we do make. There are currently adversarial models that employ architectures like XceptionNet to further improve faked content. It is this cyclic nature of this battle of technologies that makes generative modeling exciting, but also frightening.

There may come a time in the not so distant future where being able to identify fake content will be all but impossible, where generators employ all manner of counter models like XceptionNet to make the synthetic images even better than reality. There may be a point in time where we cannot ascertain what is a real image and what is fake.

When that day comes, and it likely will, how the world reacts to messages from leaders, the news, or just pictures on the internet will change forever. Will we be able to ever trust a video or image as being real? How will that affect mediums that heavily rely on image content such as photographic identification, legal proceedings, and other institutions?

Indeed, we are on the cusp of a new revolution in our digital, and to a certain extent our physical, world. This world, which was once dictated by digital and virtual media, might have to go back to relying on physical interaction, where it becomes more important to physically witness a concert or political event in person, over trusting news or other media.

As you may realize now, the deepfakes arena is only a small subset of the possibilities available to us with generative modeling. It will likely remain in the forefront and dictate the how and why of the growth and use of generators. This is not something that is likely to change until other generator applications can show they are useful in the mainstream.

I hope that through your journey in this book you have picked up the skills and understanding of how generators can be used in a variety of applications. While most of these applications will currently be centered around images or imagery like mapping or art/design, GM will certainly not be limited to just creating synthetic image content.

"We are made wise not by the recollection of our past, but by the responsibility for our future." —*George Bernard Shaw*

This great quote by George Bernard Shaw should help guide you on your journey to working with and developing your generators and fake content. Remember that your future is tightly tied to your responsibility for how you use and plan to use this technology. Using this technology ethically will be important to your future.

There are countless possibilities for applications of generative modeling in all industries worldwide, including perhaps music, text, and other forms of content. You now have the skills to jump into any of several areas where content generation could be used. I hope you use your newly acquired skills wisely and enjoy newfound success due to understanding a form of AI that will revolutionize our world in the coming years.

APPENDIX A

Running Google Colab Locally

Colab provides a feature whereby instead of using a cloud instance a user may connect to a local notebook instance. This can be useful for testing deployments of models or if your computer has a more powerful GPU than the one provided by Google for free. If you are frustrated that your cloud instance keeps shutting down, then another alternative is to set up the feature of saving and restoring models, which is detailed in Appendix C.

The instructions here are just a guideline as the actual steps may likely change as the Google Colab platform matures. Follow these general guidelines to get Colab running locally on your desktop computer:

1. Install Python and Pip locally on your system. Anaconda, a Python platform, is great for newcomers and experts. You can install Anaconda at Anaconda.com.

2. Set up a Python virtual environment based on your chosen platform and then activate it. Anaconda provides an excellent toolset to create and activate virtual environments. Just refer to the online documentation for the steps.

3. Install Jupyter into your virtual environment using this document:

   `https://jupyter.org/install`

4. PyTorch installation comes next; in the past, this could be troublesome on some operating systems. Your best resource is here:

   `https://pytorch.org/get-started/locally/`

© Micheal Lanham 2021
M. Lanham, *Generating a New Reality*, https://doi.org/10.1007/978-1-4842-7092-9

5. After your environment is set up, be sure to launch a local instance of Jupyter Notebook following the online documentation. You can also quickly smoke test your environment by doing standard PyTorch imports in a cell.

6. Open a Google Colab notebook in a browser and then click the Connect button in the upper-right menu bar. From there click "Connect to local runtime," as shown in Figure A-1.

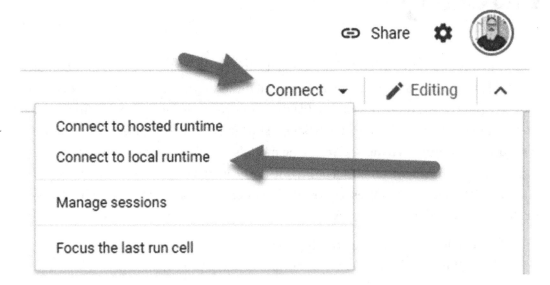

Figure A-1. *Connecting to the local runtime*

7. This will open the dialog shown in Figure A-2. Follow the instructions in the dialog, and be sure to also click the instructions link for additional setup tips.

Local connection settings

Create a local connection by following <u>these instructions</u>.

⚠ Make sure you trust the authors of this notebook before executing it. With a local connection, the code you execute can read, write, and delete files on your computer.

⚠ By default, all code cell outputs are stored in Google Drive. If your local connection will access sensitive data and you would like to omit code cell outputs, check the option below or under "Notebook settings".

☐ Omit code cell output when saving this notebook

Backend URL e.g. http://localhost:8888/?token=abc123
//localhost:8888/?token=4a8f18aa8ea8b12216d5c8232298f4f4a88cfbe050e94ec1

Start a Jupyter notebook server on your machine. Once the server has started, it will print a message with the initial backend URL used for authentication. Copy this URL in its entirety.

CANC__   CONNECT

Figure A-2. *Creating a local connection*

8. You will need to launch a new instance of Jupyter Notebook locally with the provided URL. When Jupyter is running locally, you can click the Connect button in Colab.

9. There are numerous possible reasons these simple steps could fail. If they do, carefully review the steps you undertook to install each component.

If you do get an instance of Jupyter Notebook running locally with PyTorch but still have issues connecting with Colab, an alternative is to just download the notebooks. Using the menu in Colab, select File ➤ Download .ipynb to your local machine. Then you can open the downloaded notebook directly in Jupyter on your local machine.

Generally, you won't need to run the notebooks offline, but if you do, these instructions represent a couple of options.

APPENDIX B

Opening a Notebook

Accessing and running the notebooks from the online GitHub repository can be accomplished with a few clicks, as outlined here:

1. Install Chrome if needed. Google Colab typically functions best with Chrome and will not run at all in Edge or other browsers.

2. Open Google Colab in your browser at `https://colab.research.google.com/`.

3. From the menu, select File ➤ Open notebook.

4. After the dialog opens, enter the text for the source repository URL, **`https://github.com/cxbxmxcx/GenReality`**, on the GitHub tab, as shown in Figure B-1.

© Micheal Lanham 2021
M. Lanham, *Generating a New Reality*, https://doi.org/10.1007/978-1-4842-7092-9

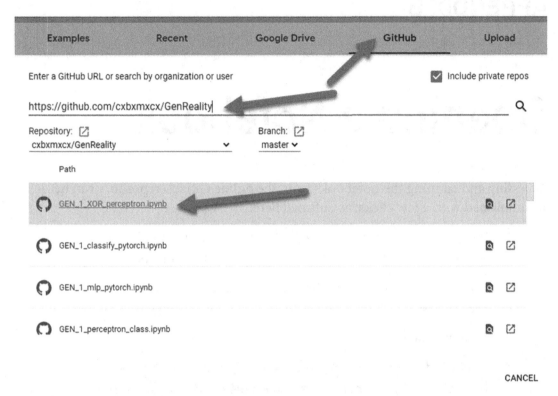

Figure B-1. *Entering the URL*

5. Locate the notebook you are looking to open from the list in the repository and click the link.

6. This will open the selected notebook, and you will be good to proceed with the exercise.

There are many other options for opening and loading notebooks into Colab that are well documented in other places.

APPENDIX C

Connecting Google Drive and Saving

Often if you are running long-running training sessions, you will want to be able to save the state of the session periodically if Colab resets. Google suggests that a session is typically good for 12 hours, but this should be considered a maximum only. You may often encounter resets that occur more frequently.

Warning   Saving state relies on accessing your Google Drive. The free version you started with provides 15 GB of storage, which sounds like a lot. However, if you are saving any number of models and output, this can quickly get eaten up and cause your Google Drive to freeze, preventing you from saving and even limiting your Gmail account and access to online Google documents.

It is therefore recommended to use saving only for single exercises, like those in Chapter 8, or if you have an extended Google Drive account. Extended Drive accounts start at less than $5 a month for 100 GB of storage. Just be aware that it is still easy to fill up 100 GB when training complex models.

The instructions in this appendix refer to accessing your Google Drive from Colab and saving state for the advanced generators covered in Chapter 8. Adding the ability to save state to the other notebooks is not difficult but requires file path changes as well as coding to save the ongoing model state. There are plenty of examples and documentation online for saving the PyTorch model state.

Follow these instructions to connect to your Google Drive from Colab. This will allow you to continue running configured notebooks even after a reset. Again, this is a great feature for the more advanced generators we covered in Chapter 8.

309

M. Lanham, *Generating a New Reality*, https://doi.org/10.1007/978-1-4842-7092-9

1. The code to mount your Google Drive to Colab is shown here:

```
from google.colab import drive

drive.mount('/content/gdrive')
```

2. Running that code from a notebook cell will generate output
    similar to Figure C-1.

Figure C-1. *Output*

3. Clicking the link will open a new browser tab requiring you to log
    into a Google account. Follow the instructions to log in to your
    account.

4. After you complete logging in, you will be prompted with a code,
    as shown in Figure C-2.

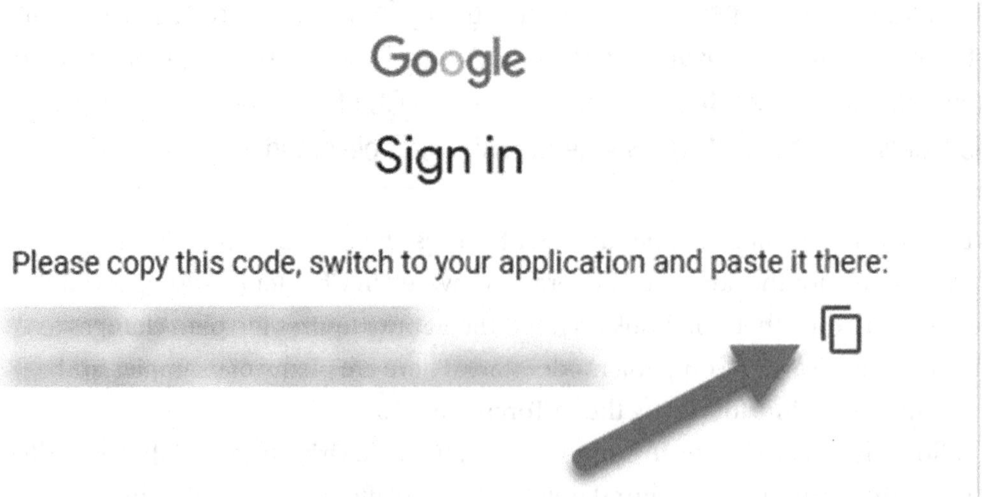

Google

Sign in

Please copy this code, switch to your application and paste it there:

Figure C-2. *Code*

5. Click the document icon to copy the code to your clipboard.

6. Return to your notebook and paste the code using Ctrl+V or Command-V into the Authorization Code field.

7. After you paste the code, you will now be able to access files in your Google Drive through the file explorer on the left, shown in Figure C-3.

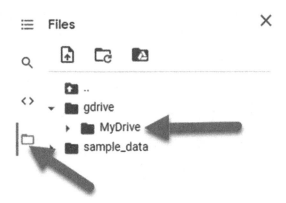

Figure C-3. *File explorer*

8. If you cannot see the drive after connecting, be sure to refresh the view or check that you copied the code correctly.

There are plenty of ways in which saving state or the data you download can save time reloading or resetting the training. However, it is important to be aware that the amount of data and the size of models you may save can be large. If you are using a free Google Drive account, then you will want to monitor this closely.

If your Google Drive does run out of space, it may prevent you from receiving email to Gmail or other critical tasks. You can clean out the space on your Google Drive by going to the account page in your browser at `https://drive.google.com/drive/my-drive`.

From that page, you can download and delete files if you need to clean up space. Be aware that when you dump files to the trash, they are not permanently deleted. To delete files and clean up drive space, you need to remove any unwanted files from the Trash folder as well.

Index

A

Activation function, 12, 19
Adaptive instance normalization (AdaIN)
    mapping network, 232
    mixing styles, 233
    removing stochastic/traditional
        input, 233
    shifting/scaling model, 232
    stochastic variation (noisy input), 233
    StyleGAN architecture, 232
    truncation, 234
    truncation (wavg w), 234
Artificial intelligence (AI), 69
ArtLine
    definition, 249
    FeatureLoss class, 250
    images, 251
    requirements.txt file, 250
    source code, 251
    training output, 252
Attention
    augmented convolution, 205–208
    convolutional layers, 199
    definition, 195
    language translation, 198
    Lipschitz continuity, 209–214
    meaning, 196
    mechanisms, 201
    models, 196
    mouse movement, 196, 197
    multihead attention, 203
        apply/multiply, 204
        random input/weights, 203
        softmax function, 204
    NLP model, 199
    SAGAN (*see* Self-attention GAN
        (SAGAN))
    Seq2Seq models, 197, 198
    transformer architecture, 202
    types, 199–201
Autoencoder
    architecture, 37
    concept learning, 36
    DataLoader, 38
    embedding, 43
    encoding/decoding, 36
    fashion dataset, 37–43
    MNIST dataset, 39
    torchvision.utils module, 38
    training process, 42

B

Backpropagation, 16, 17
Batch gradient descent (BGD), 18
BicycleGAN (Riding latent space)
    architecture, 156
    definition, 156
    Hyperparameters, 157
    models/optimizers, 158
    encoder model, 157